Harold Medway Martin

Statically Indeterminate Structures and the Principle of Least

Work

Harold Medway Martin

Statically Indeterminate Structures and the Principle of Least Work

ISBN/EAN: 9783743377288

Manufactured in Europe, USA, Canada, Australia, Japa

Cover: Foto ©berggeist007 / pixelio.de

Manufactured and distributed by brebook publishing software
(www.brebook.com)

Harold Medway Martin

Statically Indeterminate Structures and the Principle of Least

Work

STATICALLY

INDETERMINATE STRUCTURES

AND THE

PRINCIPLE OF LEAST WORK.

BY

HAROLD MEDWAY MARTIN, WH. Sc.

Revised and Reprinted from "ENGINEERING."

LONDON :

OFFICES OF " ENGINEERING," 35 AND 36, BEDFORD STREET, STRAND, W.C.

1895.

STATICALLY INDETERMINATE STRUCTURES

AND THE

PRINCIPLE OF LEAST WORK.

INTRODUCTION.

THE subject of statically indeterminate structures has been much neglected in English engineering text-books. Special cases, such as stiffened suspension bridges and rigid arches, have, it is true, been dealt with by more than one writer, but the general question has only been considered in a few scattered papers by a number of different authors. An excellent treatise on the subject was published in Italy by Castigliano in 1879. In this volume the principle of least work was enunciated and its application to determining the stresses in a structure containing superfluous bars explained. These stresses can, it is true, be determined by other methods which Maxwell seems to have originated, but the method of least work has great practical advantages, and will be adopted as the basis of what follows.

Every metallic or wooden structure is elastic, and constitutes a spring. If a spring is loaded by a weight, it elongates, and a certain amount of work is done in this elongation. This work is stored in the spring in the form of potential energy, and can be reconverted into mechanical work, as is commonly done in clocks and watches. The stiffer the spring the less it is deformed by a given weight, and hence less work is stored in a stiff spring loaded with a 1-lb. weight than in a light one loaded by the same weight. Thus if 1 ton is hung from a steel bar of 2 square inches in section, less work is done in deforming the bar than if it was hung on a steel bar of the same length and of 1 square inch section. If a weight lies on a platform supported by four legs of elastic material, work will be done in deforming the platform and in compressing the legs.

If there had been only three legs, the ordinary principles of statics would suffice to determine the weight taken by each leg, which is then quite independent of the comparative stiffness of the legs and the platform. When, however, we have more than three legs, these statical principles no longer suffice, and to determine how much of the weight is carried by each leg it is necessary to introduce other considerations. The one great principle to which such problems can be reduced is known in dynamics as that of least action, and in such problems as we have before us as that of "least work." That is to say that the work stored in an elastic system in stable equilibrium is always the smallest possible.

As an illustration, suppose a piece of steel ribbon is bent between two stops A and B, Fig. 1, then there are several possible positions of equilibrium for such a spring. For example, the S-shaped curve A D E B is one possible position of equilibrium, and the three-lobed curve A F G B is another, whilst a third possible position of equilibrium is the single-lobed curve A C B. Of these different possible positions of equilibrium some are more stable than others. Thus a slight shock or displacement will cause either the three-lobed curve or the two-lobed one to pass into a single-lobed one like A C B, but this latter cannot by a small shock be made to pass into either of the other forms.

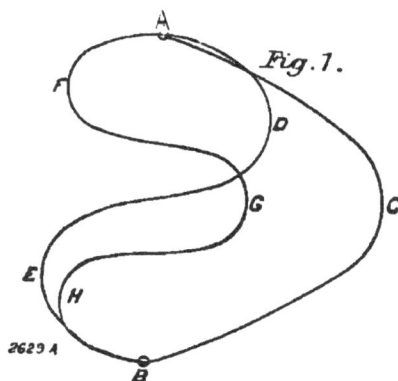

Fig. 1.

Hence it is a case of stable equilibrium whilst the others are unstable. It will be noted that the stable form is the least bent one, and as the work stored in a flat spring increases with the amount of bending, it is obvious that the work stored up as potential energy is less in the stable than in the unstable conditions of equilibrium.

Returning now to our four-legged table, there are an infinite number of ways in which the load could be distributed over the four legs, whilst still maintaining equilibrium. But if the equilibrium is to be stable, the distribution will be such that the worked stored up as potential energy will be as small as possible.

To fix the ideas, let Fig. 2 represent a table 50 in. square, supported by four legs of the same material and of uniform section, and let a weight W be placed as indicated. Numbering the legs 1, 2, 3, and 4, the load taken by each has to be ascertained. If

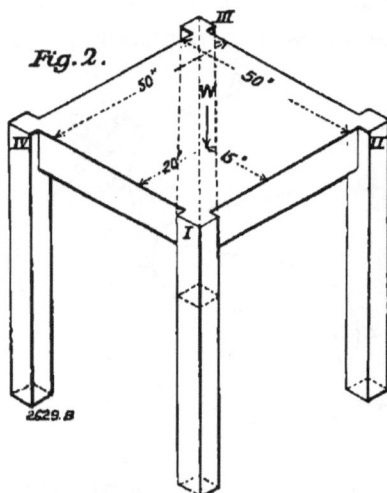

Fig. 2.

the top of the table is very rigid, compared with the legs, the problem is very easily solved by the method of least work, and we will assume this to be the case. Under these conditions, the work done in deforming the top of the table will be very small compared with that done in compressing the legs, and may, therefore, be neglected without materially affecting the results of the calculation. Now, in an ordinary spring balance, the extension or compression is twice as great for a load of 2 lb. as it is for one of 1 lb., and for P lb. the extension of the spring is P times as great as it is for 1 lb. Suppose the balance at zero, then, by placing small shot in the scale pan, we can gradually bring up the load in the pan to any desired amount P. Suppose the spring extends $\frac{1}{8}$ in. for a weight of 1 lb., then the final extension of the spring is P times this, or $\frac{P}{8}$ in. During the process of adding the shot, the load in the pan has been gradually increased from 0 to P, and hence the average weight has been $\frac{P}{2}$. Now the work done by a given force is found by multiplying the average value of the force by the distance through which it has moved. The average force in the present case is $\frac{P}{2}$ lb., and the distance it has moved is $\frac{P}{8}$ in. Hence the work stored in the spring is $\frac{P^2}{16}$ inch-pounds.

Now, coming back to our table, suppose under a load of 1 lb. each leg is compressed a in. Under a load P lb. it will then compress P a in., and the total work done in the compression will, as shown above, be $\frac{P^2 a}{2}$ inch-pounds.

Let us assume that the weight taken by leg No. 4 is P_4, and that P_1, P_2, and P_3 represent the weight taken by each of the other legs.

Evidently

$$\cdot \ P_1 + P_2 + P_3 + P_4 = W \ \ . \ \ . \ \ . \ \ . \ \ . \ \ (1).$$

Also, by taking moments, we have

$$50 (P_4 + P_3) = 15. \, W,$$

or

$$P_3 = \frac{15 \, W - 50 \, P_4}{50} = \frac{3 \, W - 10 \, P_4}{10} \quad \cdots \quad (2).$$

Similarly

$$50 (P_1 + P_4) = 30 \, W,$$

or

$$P_1 = \frac{6 \, W - 10 \, P_4}{10} \quad \cdots \cdots \quad (3)$$

$$\therefore \text{ from (1) } P_2 = \frac{W + 10 \, P_4}{10}.$$

Hence, if P_4 was known, all the others could be determined.   To determine P_4, consider the work done in compressing the legs.   Call this U, then

$$U = \frac{a}{2} \left(P_1{}^2 + P_2{}^2 + P_3{}^2 + P_4{}^2 \right),$$

and substituting for P_1, P_2, P_3, we find

$$U = \frac{a}{200} \cdot \left(46 \, W^2 - 160 \, W \, P_4 + 400 \, P_4{}^2 \right),$$

and we must choose P_4 so that this is as small as possible.   One way of doing this would be to choose a number of values for P_4, taking it successively equal to $\frac{1}{2}$ W, $\frac{1}{3}$ W, $\frac{1}{4}$ W, $\frac{1}{5}$ W, $\frac{1}{6}$ W, and so on, and to tabulate the values then found for the quantity inside the bracket.   We thus get:

when $P_4 = \dfrac{W}{2}$ then $46 \, W^2 - 160 \, W \, P_4 + 400 \, P_4{}^2 = 66 \quad W^2$

$\qquad P_4 = \dfrac{W}{3}$,, ,, ,, $= 37.1 \; W^2$

$\qquad P_4 = \dfrac{W}{4}$,, ,, ,, $= 31 \quad W^2$

$\qquad P_4 = \dfrac{W}{5}$,, ,, ,, $= 30 \quad W^2$

$\qquad P_4 = \dfrac{W}{6}$,, ,, ,, $= 30.4 \; W^2$

Now this last value is greater than that immediately preceding it, and thus after first steadily diminishing the value begins to increase again. Hence the proper value of P_4 is somewhere between $\dfrac{W}{4}$, and $\dfrac{W}{6}$, and if the numbers were plotted down, and a smooth curve drawn through them, it would be found that $\dfrac{W}{5}$ was about the correct value. This method, though quite practicable, and not difficult of application in the present instance, would in many cases prove exceedingly troublesome, and it is best to make use of the differential calculus, which gives the correct value for P_4 at once, and the differentiations involved are of the very simplest kind.

We have

$$U = \frac{a}{200}(46\,W^2 - 160\,W\,P_4 + 400\,P_4{}^2).$$

Whence

$$\frac{dU}{dP_4} = \frac{a}{200}\left\{\ -160\,W + 800\,P_4\ \right\}$$

and this is equal to 0 when U has its smallest possible value;

$$\therefore\ 800\,P_4 - 160\,W = 0,$$

or

$$P_4 = \frac{W}{5}$$

Having got this, the other reactions are found to be as follows :

$$P_2 = \frac{W + 2\,W}{10} = \frac{3}{10}\,W$$

$$P_1 = \frac{6\,W - 2\,W}{10} = \frac{4}{10}\,W$$

$$P_3 = \frac{W}{10}.$$

This, then, is the distribution of the weight on the four legs when the latter are of uniform section and material. Suppose, however, the legs were of different lengths, and had different sectional areas. In this case the first thing to be done is to determine how much each leg will be compressed by a unit force.

If l be the length of leg, Ω its area, and P the load on it, the extension will be

$$\lambda = \frac{P}{\Omega} \cdot \frac{l}{E}.$$

The work done in extending it will be, as already shown,

$$\frac{P\lambda}{2} = \frac{P^2}{2} \cdot \frac{l}{\Omega . E}.$$

Now, returning to our table, suppose legs 1 and 2 are 30 in. long and 2 square inches in area, whilst legs 3 and 4 are 20 in. long and 1 square inch in area. Then U, the work done in compressing the legs, will be :

$$U = \frac{P_1^2}{2E} \cdot \frac{30}{2} + \frac{P_2^2}{2E} \cdot \frac{30}{2} + \frac{P_3^2}{2E} \cdot \frac{20}{1} + \frac{P_4^2}{2E} \cdot \frac{20}{1}$$

$$= \frac{1}{2E} \left\{ 15 P_1^2 + 15 P_2^2 + 20 P_3^2 + 20 P_4^2 \right\}$$

and, as before, we have

$$P_1 = \frac{6 W - 10 P_4}{10}$$

$$P_2 = \frac{W + 10 P_4}{10}$$

$$P_3 = \frac{3 W - 10 P_4}{10}.$$

Whence

$$U = \frac{1}{200\,E} \left\{ \begin{array}{l} 15\,(6\,W - 10\,P_4)^2 + 15\,(W + 10\,P_4)^2 \\ + 20\,(3\,W \quad 10\,P_4)^2 + 2000\,P_4^2 \end{array} \right\}$$

$$\therefore\ 100\frac{d\,U}{d\,P_4}\,E = 0 = -900\,W + 1500\,P_4 + 150\,W + 1500\,P$$
$$-\ 600\,W + 2000\,P_4 + 2000\,P_4$$

Whence $P_4 = \frac{27}{140}\,W$, and hence $P_1 = \frac{57}{140}\,W$,

$$P_2 = \frac{41}{140}\,W, \text{ and } P_3 = \frac{15}{140}\,W.$$

LATTICE GIRDERS.

As another case of a structure with one super-fluous member, take the truss represented in Fig. 3, which represents diagrammatically one-half of a lat-tice girder erected some years back to carry a single

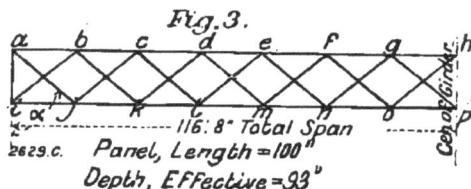

line of railway over a river in Scotland. The bridge is a deck structure, the floor resting on top of the girders. The stresses in such a structure cannot be determined by pure statics, as it contains one superfluous bar. That this is so is most easily shown by constructing a model. If, however, either the bar $a\ i$ or its fellow at the opposite abutment were removed, the stresses in all the bars could then be found by the ordinary methods. It is usual in dealing with this form of truss to assume that the stress on the bar $a\ i$ is due entirely to the loads applied at the points *c, e, g,* &c., whilst those applied at *b, d, f, h,* &c., are held to have no influence on the stress in this bar. It will be of interest to see how far these assumptions may be wrong, and we will determine the true value of the stress in *a i,* and hence in the rest of the structure, by the principle of least work. In the first place, it will be necessary to tabulate the values of $\dfrac{L}{\Omega}$ for every bar of the structure, where L denotes the length of any bar and Ω its area. For this work a small slide rule is very convenient, being quite accurate enough for the purpose, and it has been almost exclusively used in performing the various calculations of this volume. From the drawings of the structure we get the values tabulated below. The bar *a i* is omitted because its area is very great, and the work done in deforming it will be small compared with that done in deforming the rest of the structure.

Bar.	$\dfrac{L}{\Omega}$	Bar.	$\dfrac{L}{\Omega}$
a j	12.27	*a b*	3.33
j c	13.03	*b c*	3.33
c l	15.61	*c d*	2.56
l e	16.05	*d e*	2.08
e n	22.75	*e f*	1.75
n g	25.50	*f g*	1.75
g p	19.71	*g h*	1.75
i b	13.03	*i j*	3.33
b k	13.65	*j k*	3.33
k d	16.05	*k l*	2.56
d m	18.02	*l m*	2.08
m f	16.05	*m n*	1.75
f o	25.50	*n o*	1.75
o h	25.50	*o p*	1.75

The bars on the other side of the mid span will have corresponding values of $\frac{L}{\Omega}$. In practice, however, the bracing bars only require to be taken into consideration as the work done on the top and bottom flanges varies but little with different distributions of load between the two systems of web members, and hence these flanges can and will be entirely neglected in what follows:

Let u be the work done in deforming one bar, then the whole work will be equal to the work done in deforming all the bars, or

$$U = \Sigma u.$$

But if $\mu =$ the stress in the supernumerary bar $a\,i$, the work done in deforming the whole structure will be a minimum with respect to μ. Whence

$$\frac{d\,U}{d\,\mu} = 0 = \Sigma \frac{d\,u}{d\,\mu},$$

but for any bar, as has already been shown above,

$$u = \frac{T^2 L}{2\,E\,\Omega},$$

where T is equal to the stress in the bar, and T can always be expressed in the terms of the external load on the structure and μ. Whence $\Sigma \frac{d\,u}{d\,\mu}$, can be found, and equating this to zero we shall get the value of μ.

Suppose, for example, the central apex h only is loaded by a weight W. Let R be the reaction on the left-hand abutment, and μ the stress in the bar $a\,i$. Then in this case from symmetry it will be obvious that the stress in the bar corresponding to

$a\,i$ at the right hand abutment, will also be μ. But on drawing a stress diagram it will be found that for equilibrium the stress in this bar should be $-\mu$. Hence in the case taken $\mu = o$. and the load is carried wholly by one system of bracing although if any bar of this system were removed the structure would still support the load the other system then coming into play. If, however, the load was placed at the apex f instead of at h, then a certain amount of stress will obtain in both systems of bracing. As before let μ be stress in the bar $a\,i$. Then the shear causing stress in the bars $i\,b,\,b\,k$, &c., will be $R - \mu$ where R is the total reaction at the left abutment. The shear causing stress in the bars $a\,j,\,j\,c$ will of course be μ.

Hence the stress in the bars $i\,b,\,b\,k$, &c., will be $\pm\,(R - \mu)\sec a$ up to the apex f, and beyond this point for the remaining bars of this system it will be $\pm\,\{\,R - \mu - W)\sec a$. For the other system the stress in the bracing bars will be $\overline{+}\,\mu\sec a$ throughout, a being the angle between the bars and the vertical.

Hence for each of the bracing bars of the first system up to the apex f,

$$u = \frac{L}{2\,E\,\Omega}\sec^2 a\,(R - \mu)^2$$

and beyond this point $u = \dfrac{L}{2\,E\,\Omega}\cdot\sec^2 a\,(R - \mu - W)^2$

$$\therefore \frac{d\,u}{d\,\mu} = \frac{\sec^2 a}{E}\cdot[\mu - R]\frac{L}{\Omega}\cdot \text{ up to } f,$$

$$\text{and} = \frac{\sec^2 a}{E}\,[\mu + W - R]\frac{L}{\Omega}$$

for the rest of the bars of the system. Hence taking the value of $\frac{L}{\Omega}$. from the proceeding table we get for the first system of bracing,

$$\frac{se^2\,a}{E}\,2\left\{\,\mu-R\,\right\}\left\{\,13.03+13.65+16.05+18.02+16.05+\right.$$

$$25\,50+25.50\,\Big\}+\frac{sec^2\,a}{E}\cdot W\left\{\,25.50+25.50+25.50+25.50+\right.$$

$$16.05+18.02+16.05+13.65+13.03\,\Big\}$$

or

$$\frac{sec^2\,a}{E}\cdot\left\{\,255.6\,\mu-255.6\,R+178.8\,W\,\right\}.$$

And taking moments, about the right-hand abutment we get, $R=\frac{9}{14}\cdot W$. Whence on substituting, there results for the sum of $\frac{d\,u}{d\,\mu}$, for this first system of bracing,

$$\frac{sec^2\,a}{E}\left\{\,255.6\,\mu-15.6\,W\,\right\}.$$

Similarly for the second system of bracing we have throughout for each bar

$$\frac{d\,u}{d\,\mu}=\frac{sec^2\,a}{E}\cdot\mu\cdot\frac{L}{\Omega},$$

and substituting from the table as before and adding we get

$$249.8\,\frac{sec^2\,a}{E}\cdot\mu$$

Hence putting $U=\Sigma\,u$ we get

$$\frac{d\,U}{d\,\mu}=\frac{sec^2\,a}{E}\left\{\,255.6\,\mu+249.8\,\mu-15.6\,W\,\right\}=0,$$

when U is a mininum.

$$\therefore\,505.4\,\mu=15.6\,W.$$
$$\text{whence }\mu=.031\,W.$$

The value of μ for other portions of the load will be as follows :

Load at					Value of μ.
b	...	...	...	...	...	.011 W.
d	...	...	...	...	...	.024 W.
f	...	...	...	...	...	.031 W.
h	...	...	...	...	...	.000 W.
c	...	...	...	...	...	.877 W.
e	...	...	...	...	...	.743 W.
g	...	...	...	...	...	.575 W.

If any three apices such as b, c and d were loaded, each with a load W. the resulting stress in the bar ai. will of course be got by summation, and will in the case given be W $(\cdot011 + \cdot877 + \cdot024) = \cdot912$ W. Knowing this the stress in any other bar can be determined by pure statics. In the girder considered the modification in the stresses as obtained on the usual assumption is trivial. The example is of interest, however, as many American Engineers object to the use of double systems of triangulation on the ground that the stresses in them cannot, strictly speaking, be determined by pure statics, the usual assumptions being objected to as being erroneous. From the complete investigation above given, it would appear that the error involved in these assumptions may well be neglected in practice.

TRUSSES WITH MORE THAN ONE SUPERFLUOUS BAR.

A lattice girder of the type dealt with has one superfluous member only, but the method of least work is also applicable to structures containing more than one such member. As an instance, take

the truss shown in Fig. 4, which, though of a type
not to be recommended for adoption in an actual
bridge, will, nevertheless, afford a good example
of the methods to be employed in calculating such
a framework. This truss contains two superfluous
members. Thus, if the bars B E and E G were removed, the stresses in the various members could
be determined by ordinary statics. But these
stresses could also be determined if the stresses
in these two bars were known. Thus let the stress
in B E be P, and in E G be Q. Assume each
of these bars is in tension, then their action on
the structure will be the same as if the bars were

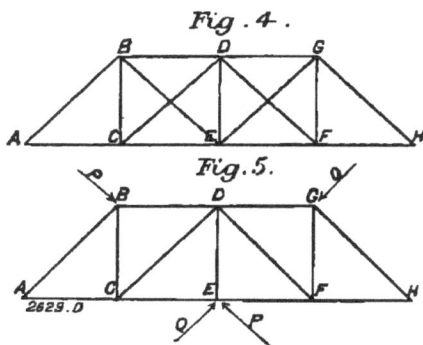

Fig. 4.

Fig. 5.

taken away altogether and forces applied to the joints
B, G, and E, as indicated in Fig. 5. This latter figure
represents, it will be seen, a statically determined
structure, and the stresses in all the bars can be expressed in terms of the loading and P and Q, whilst
these latter can be determined by the method of least
work. To apply this principle, however, it is necessary
to know the ratio of $\frac{L}{\Omega}$ for each bar, where L = length

of bar and Ω its area. These ratios can be determined with sufficient accuracy by the ordinary methods of dealing with such structures, viz., by dividing them up into two separate trusses, each of which is statically determined, and finally combining the results. An approximation to the true stress is thus obtained, permitting the section of the bars to be designed provisionally, and the ratio $\frac{L}{\Omega}$ found. Suppose the ratios determined in this way to be those given in the second column of the annexed Table. Then the third column shows the stresses due to a load of 29 tons at C, the rest of the structure being unloaded. A small slide rule has been used for these calculations, and is quite accurate enough for the purpose. The total stress, obtained by adding up columns 3, 4, and 5, is given in column 6. Then let u be the work done in deforming any bar, B D, for example. Then

$$u = \frac{L}{2E\Omega} \cdot \left\{ .707\,P + 21.8 \right\}^{2}.$$

Hence

$$\frac{du}{dP} = \frac{1}{E} \cdot \frac{L}{\Omega} \cdot \left\{ .707\,P + 21.8 \right\} \times .707,$$

and

$$\frac{L}{\Omega} \text{ for bar B D} = 13.5.$$

$$\therefore \frac{du}{dP} = 6.75\,P + 208.$$

In this way columns 7 and 8 are obtained, and finally, if U is the whole work done in deforming the structure, $U = \Sigma u$; and hence

$$\frac{dU}{dP} = \Sigma \frac{du}{dP}, \text{and} \frac{dU}{dQ} = \Sigma \frac{du}{dQ}.$$

TABLE I.

STRESSES IN MEMBERS OF TRUSS, FIG. 4.

1. Bar.	2. $\dfrac{L}{D}$	3. Stress Due to a Load of 29 Tons at C.	4. Stress Due to P.	5. Stress Due to Q.	6. Total Stress.	7. $\dfrac{du}{dP} = \dfrac{1}{E} \times$	8. $\dfrac{du}{dQ} = \dfrac{1}{E} \times$
A B		$+30.7$			$+30.7$	$14.75\,P - 455$	$34.65\,(P+Q)$
A C		-21.8			-21.8	$6.75\,P + 208$	$11.1\,Q - 228$
C B	29.5	$+21.8$	$+.707\,P$		$+.707\,P - 21.8$	$30.\,3\,P + 309$	$30.\,3\,Q - 309$
B D	13.5	-21.8	$+.707\,P$		$+.707\,P + 21.8$	$11.\,1\,P - 228$	$6.75\,Q + 69$
D C	30.3	-10.2	$-P$		$-10.2 - P$	$34.65\,(P+Q)$	$14.75\,Q - 152$
C E	22.2	-14.5	$+.707\,P$		$+.707\,P - 14.5$		
D E	69.3	0	$+.707\,P$	$+.707\,Q$	$+.707\,P + .707\,Q$		
F E	22.2	-14.5		$+.707\,Q$	$+.707\,Q - 14.5$		
D F	30.3	$+10.2$		$-Q$	$+10.2 - Q$		
G D	13.5	$+7.3$		$+.707\,Q$	$+.707\,Q + 7.3$		
F G	29.5	-7.3		$+.707\,Q$	$+.707\,Q - 7.3$		
H F		-7.3			-7.3		
H G		$+10.2$			$+10.2$		
B E	85		$-P$		$-P$	$85\,P$	
E G	85			$-Q$	$-Q$		$85\,Q$

But for a minimum

$$\frac{d\,U}{d\,P} = 0, \text{ and } \frac{d\,U}{d\,Q} = 0.$$

Hence, adding up columns 7 and 8 we get

$$\frac{d\,U}{d\,P} = 0 = 182.55\,P + 34.65\,Q - 166,$$

and

$$\frac{d\,U}{d\,Q} = 0 = 34.65\,P + 182.55\,Q - 620.$$

Whence P = .27 tons, and Q = 3.35 tons. Then the stress in bar B E is − .28 tons, and in bar E G is − 3.35 tons.

By a precisely similar process, the stresses can be computed, at least theoretically, in a truss containing any number of superfluous bars, but when the number of these bars becomes large, the labour of computation becomes excessive. The theory of the process is, however, as shown in the preceding examples, very simple. The truss is redrawn with the superfluous bars removed, and the stresses in the statically determined structure, calculated by ordinary statics; unknown forces P, Q, R, &c., being substituted to represent the action of the bars removed. The values $\frac{L}{\Omega}$ are then tabulated for every bar and from these the values of

$$\frac{d\,U}{d\,P}, \frac{d\,U}{d\,Q}, \frac{d\,U}{d\,R}, \text{ &c.,}$$

are determined and equated to zero. By solving the simultaneous equations thus obtained, the values of P, Q, R, &c., can be found, and once these are known the stress in every other bar of the structure can be obtained by ordinary statics.

DEFLECTION.

The opinion is very commonly held that in calculating the deflection of a bridge truss, the deflection due to the strains of the bracing bars can be neglected, and that it is sufficient to take into account the extension and compressions of the booms only. Whilst this is true for very shallow girders, it is not so for the deep trusses used in American practice, and it will now be shown how the deflection of such a truss can be calculated from the elastic work of deformation. We shall first prove the following theorems:

The deflection of an elastic structure at any point is equal to the differential coefficient of the work done in deforming the structure with respect to the load acting at that point. Thus, if Δ be the deflection of a beam at a certain point b in it, W_b a load applied there, and U the total work of deformation, then

$$\Delta = \frac{d\,U}{d\,W_b}\,.$$

For, suppose a truss loaded at the points a, b, c, and d, &c., with loads W_a, W_b, W_d, &c. First assume all these loads to be zero, and let them be uniformly increased up to the values W_a, W_b, W_c, &c., and let the final deflections at each of the points be Δ_a, Δ_b, Δ_c, &c. Then the load at a has moved through the distance Δ_a, and has increased uniformly from 0 to W_a. Hence its average value has been $\frac{W_a}{2}$. Multiplying this average value of the force by the distance moved through, the work done by it on the beam has been $\frac{W_a}{2} \cdot \Delta_a$, and similarly for b, c, d, &c.

Hence the total work,

$$U = \frac{W_a \, \Delta_a}{2} + \frac{W_b \, \Delta_b}{2} + \frac{W_c \, \Delta_c}{2} + , \&c.$$

Suppose now the force W_a is increased by a very small quantity $d \, W_a$. Then the deflection at each of the points considered will also be increased by correspondingly small quantities. Thus the deflection at b will be increased by an amount $\frac{\delta \, \Delta_b}{\delta \, W_a} . \, d \, W_a$, and so on. Hence the total work U will be increased by a quantity

$$d \, U = [W_a + d \, W_a] \frac{\delta \, \Delta_a}{\delta \, W_a} . \, d \, W_a + W_b \frac{\delta \, \Delta_b}{\delta \, W_a} . \, d \, W_a$$

$$+ \, W_c \frac{\delta \, \Delta_c}{\delta \, W_a} . \, d \, W_a + , \&c.$$

Then, neglecting the squares of small quantities and dividing by $d \, W_a$, we get

$$\frac{d \, U}{d \, W_a} = W_a \frac{\delta \, \Delta_a}{\delta \, W_a} + W_b \frac{\delta \, \Delta_b}{\delta \, W_a} + \, W_c \frac{\delta \, \Delta_c}{\delta \, W_a} + , \&c. \quad (1)$$

But, as already shown.

$$2 \, U = W_a \, \Delta_a + W_b \, \Delta_b + \&c.$$

$$\therefore 2 \frac{d \, U}{d \, W_a} = \Delta_a + W_a \frac{\delta \, \Delta_a}{\delta \, W_a} + W_b \frac{\delta \, \Delta_b}{\delta \, W_a} + , \&c. \quad (2)$$

Then substracting (1) and (2) we get finally

$$\Delta_a = \frac{d \, U}{d \, W_a},$$

and similarly

$$\Delta_b = \frac{d \, U}{d \, W_b}, \quad \Delta_c = \frac{d \, U}{d \, W_c}, \, \&c.$$

Hence, finally, the deflection at any point of an elastic structure is equal to the differential coefficient of the work of deformation, with respect to the load applied at that point.   Q.E.D.

The deflection at the point b of an elastic structure caused by load W at the point a is equal to the deflection at a in the direction of the original load, caused by an equal load at the point b applied in the direction of the original deflection. Suppose, as before, a beam loaded at the points a, b, c, &c.

Let the deflection at b caused by the load at a be $\beta_1 W_a$, and similarly for W_b, W_c, &c., so that the total deflection at b is

$$\Delta b = \beta_1 W_a + \beta_2 W_b + \beta_3 W_c +, \&c.,$$

and similarly deflections at a and c are respectively

$$\Delta a = a_1 W_a + a_2 W_b + a_3 W_c +, \&c.,$$
$$\Delta c = \gamma_1 W_a + \gamma_2 W_b + \gamma_3 W_c +, \&c.$$

Then let U be the work done in deforming the structure. Then it has been shown that

$$U = \tfrac{1}{2}[W_a\ \Delta a + W_b\ \Delta b + W_c\ \Delta c +, \&c.]$$

or

$$U = \frac{W_a}{2}\ [a_1\ W_a + a_2\ W_b + a_3\ W_c +, \&c.]$$

$$+ \frac{W_b}{2}[\beta_1\ W_a + \beta_2\ W_b + \beta_3\ W_c +, \&c.]$$

$$+ \&c., + \&c.$$

Then

$$\frac{d\ U}{d\ W_a} = \tfrac{1}{2}\ [a_1\ W_a + a_2\ W_b + a_3\ W_c +, \&c.]$$

$$+ \frac{W_a}{2}\ a_1 + \frac{W_b}{2}\ \beta_1 + \frac{W_c}{2}\ \gamma_1 +, \&c.$$

But it has already been shown that

$$\frac{d\ U}{d\ W_a} = \Delta a = a_1\ W_a + a_2\ W_b + a_3\ W_c +, \&c.$$

and equating these two values of

$$\frac{d\ U}{d\ W_a},$$

it is evident that

$$\beta_1 = \alpha_2, \ \gamma_1 = \alpha_3, \ \&c.$$

But the deflection caused at the point b by a load W at a is by hypothesis β_1 W, and similarly the deflection caused at the point a by a load W at b is a_2 W. But $\beta_1 = a_2$; which proves the proposition.

THE HAWKESBURY BRIDGE.

As an example of the method, let it be required to find the deflection of the truss shown in Fig. 6, when loaded with 30 tons at each panel point. The truss in question is one of those used for the

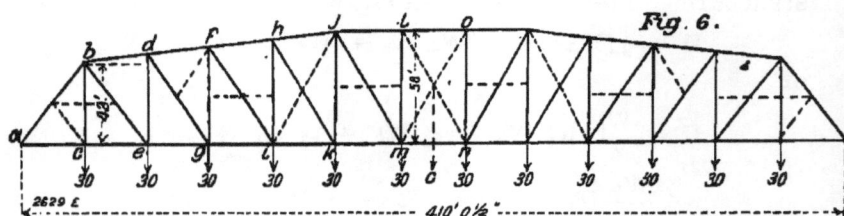

Fig. 6.

Hawkesbury Bridge, N.S.W. The bars shown dotted are counters, stiffening bars, or are otherwise unstrained under a uniform load. To determine the deflection, the length and area of each acting bar must be determined. These are given in the second and third columns of Table II., which extends up to the centre of the bridge only, the truss being symmetrical about mid span. The ratio of these $\dfrac{L}{\Omega}$ is given in the fourth column, whilst in the fifth column the stresses due to a load of 30 tons at each panel point are written down. As in previous cases,

TABLE II.

1.	2.	3.	4.	5.	6.	7.
Bar.	L	Ω	$\dfrac{L}{\Omega}$	Stress due to 30 Tons at each Panel Point.	Stress due to C.	$\dfrac{\left(\dfrac{d\,u}{d\,C}\right)_o}{\dfrac{1}{E}\times}$
	in.	sq. in.				
a b	625.2	175	3.57	+ 224.7	+ .626 C	+ 502
a c	378.5	60	6 31	− 135.0	− .376 C	+ 320
c b	504.0	16	3.15	− 30.0	0	0
c e	378.5	74	5.11	− 135.0	− .376 C	+ 259
e b	625.2	62	10.11	− 150.3	− .530 C	+ 805
b d	381.6	138	2.76	+ 228.0	+ .706 C	+ 444
d e	552.0	60.7	9.10	+ 91.5	+ .410 C	+ 341
e g	378.5	118	3.21	− 226.5	− .706 C	+ 512
d g	558.0	43.7	12.76	− 100.0	− .460 C	+ 587
d f	381.6	170	2.24	+ 294.1	+ .962 C	+ 634
g f	600.0	43.5	13.80	+ 54.0	+ .380 C	+ 283
g i	378.5	148	2.56	− 282.0	− .962 C	+ 694
f i	709.2	28.9	24.50	60.0	− .418 C	+ 614
f h	381.6	187	2.04	+ 317 1	+ 1.188 C	+ 762
i h	648.0	28	23.19	+ 19.8	+ .350 C	+ 161
i k	378.5	166	2.28	− 315.0	− 1.183 C	+ 843
k h	750.0	18	41.60	− 20.7	− .380 C	+ 327
h j	381.6	195	1.95	+ 328.5	+ 1.380 C	+ 884
j k	696.0	22	31.60	− 12.0	+ .326 C	− 125
k m	378.5	172	2.20	− 325.8	− 1.376 C	+ 941
j m	793.2	22.5	35.21	− 32.4	− .572 C	+ 652
j l	378.5	201.4	1.88	+ 341.4	+ 1.376 C	+ 884
m l	693.0	22	31.60	0	+ .500 C	0
m n	378.5	178	2.13	− 341.4	− 1.648 C	+ 1199
l n	—	—	—			
l o	378.5	201.4	1.88	+ 341.4	− 1.975 C	+ 1268

a slide rule has been used. Now to determine the

deflection. Suppose a force C to be applied at the centre of the bridge as shown. This force will give rise to the stresses given in the sixth column of the Table. But the deflection at the centre of the bridge will, as we have shown, be equal to

$$\frac{d\,U}{d\,C},$$

where U is the work done in deforming the structure, and this will be true whatever value C has. Hence to find the deflection under the load of 30 tons at each panel point, it is only necessary to determine

$$\frac{d\,U}{d\,C},$$

and then to make $C = 0$.

Let u be the work done in deforming any separate bar of the structure. Then

$$U = \Sigma\,u,$$

and

$$\frac{d\,U}{d\,C} = \Sigma\,\frac{d\,u}{d\,C}.$$

Now

$$u = \frac{1}{2\,E} \cdot \frac{L}{\Omega} \cdot S^2,$$

where S is the stress in the bar. Thus for bar $c\,e$

$$S = -\,(135 + .376\,C),$$

and

$$\frac{L}{\Omega} = 5.11.$$

Hence for this bar

$$u = \frac{5.11}{2\,E} \cdot (135 + .376\,C)^2,$$

and

$$\frac{d\,u}{d\,C} = \frac{1}{E} \cdot \left\{ \ 259 + .721\ C \ \right\}.$$

Putting $C = 0$ we get

$$\left(\frac{d\,u}{d\,C}\right)_0 = \frac{259}{E} \text{ in.,}$$

which is the deflection at the centre of the bridge due to the stress in the bar $c\ e$. The total deflection will be equal to the sum of the deflections due to every bar, *i.e.*, to

$$\Sigma \left(\frac{d\,u}{d\,C}\right)_0$$

where the suffix denotes that C has been made zero after the differentiation. The values of

$$\left(\frac{d\,u}{d\,C}\right)_0$$

are tabulated in column 7.

To find the total deflection, all values from $a\ b$ to $m\ l$ must be counted twice, as corresponding bars occur on both sides of the centre line. The values for $m\ n$ and $l\ o$ will, of course, be taken only once. In this instance only one of the terms in column 7 is negative, and taking account of this minus sign, it will be found that

$$\left[\frac{d\,U}{d\,C}\right]_0 = \Sigma\left(\frac{d\,u}{d\,C_0}\right) = \frac{25,315 \text{ in.}}{E}.$$

Hence if $E = 13,000$ tons per square inch, the deflection of the truss under the assumed loading will be

$$\Delta = \left(\frac{d\,U}{d\,C_0}\right) = \frac{25,315}{13,000} = 1.95 \text{ in.}$$

If the deflection due to the bracing had been

neglected, as is commonly the case, the calculated deflection would have been

$$\Delta = \frac{18,025}{13,000} = 1.386 \text{ in. only.}$$

It thus appears that in this particular truss the deflection due to the deformation of the bracing bars is more than 40 per cent. of that due to the deformation of the top and bottom chords only. Had the truss parallel booms, the proportion of the deflection due to the bracing would have been still greater. It should, however, be noted that the truss considered above is of considerable depth as compared with the span, and with shallow girders the effect of the bracing on the deflection is much less important.

The great advantage of the method for calculating deflections, explained above, lies in the fact, that there is never any difficulty in determining whether the deflection due to the stress in any particular bar is positive or negative, as the proper sign for each particular quantity is determined almost automatically. This is particularly useful in very complicated cases.

SWING BRIDGES.

From what has been shown above, it is evident that considerable errors may be made in determining the reactions on the piers and abutments of a swing bridge in the usual way by the theorem of three moments. Moreover, with a type of

bridge such as that shown in Fig. 7, this theorem
ceases to be applicable for unsymmetrical loads, as
there is no bracing in the central panel. The truss
in question originated in America, but our figure
represents an English example. The level of the
rocker plates at the abutments is so fixed that the
whole of the dead load is carried on the central
pier. Hence, in determining the reactions at the
abutments, the live load only has to be considered.
It may be taken as 13 tons at each panel point of
the lower chord. The abutment reaction for the
live load tends to reverse the stress in diagonals
c d, e f, y h, &c., but the dead load stresses being

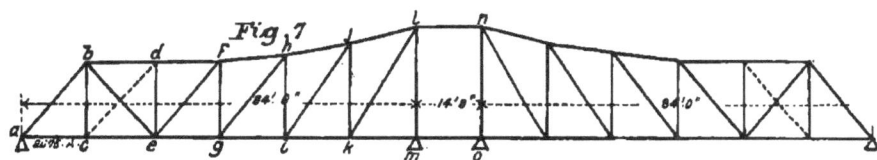

Fig. 7

high, the panel c e on each side alone requires
counterbracing.

There being no bracing in the central panel, it
will be obvious that if the bar l n were cut through
we should have two completely independent trusses,
the reactions in which could be determined by
pure statics. Assuming this done, the stresses in the
various bars will be found to be those given in the
the Table on page 30. If now the stress in l n is
assumed to be equal to P, the stresses due to this
will be those given in column 3. Then, as before,
the unknown quantity P can be determined by the
method of least work. Thus, if S is the total stress

in any bar, l its length, and Ω its area, the work done in compressing or extending this bar will, as before, be

$$u = \frac{L\,S^2}{2\,\Omega\,E},$$

TABLE III.

Bar.	$\dfrac{L}{\Omega}$	Stress due to 13 Tons at each Panel Point.	Stress Due to P.	Total Stress S.	$\dfrac{L\,S}{\Omega} \cdot \dfrac{d\,S}{d\,P}$
$a\ b$	12.19	+ 43.0	− .366 P	+ 43.0 − .366 P	1.580 P − 185.5
$a\ c$	7.38	− 28.3	+ .239 P	− 28.3 + .239 P	.422 P − 48.2
$b\ c$	13.25	− 13.0	0	− 13.0	0
$c\ c$	7.38	− 28.3	+ .239 P	− 28.3 + .239 P	·422 P − 48.2
$b\ d$	7.19	+ 45.0	− .485 P	+ 45.0 − .485 P	1.688 P − 156.6
$b\ e$	44.9	− 25.7	+ .375 P	− 25.7 + .375 P	6.300 P − 432.5
$d\ c$	13.26	+ 1.4	− .014 P	+ 1.4 − .014 P	.003 P − .3
$d\ f$	7.19	+ 45.0	− .485 P	+ 45.0 − .485 P	1.688 P − 156.6
$e\ g$	7.38	− 49.3	+ .710 P	− 49.3 + .710 P	3.714 P − 258.0
$e\ f$	22.83	+ 6.8	− .350 P	+ 6.8 − .350 P	2.795 P − 54.3
$f\ h$	11.80	+ 49.5	− .712 P	+ 49.5 − .712 P	6.975 P − 415.5
$f\ g$	13.10	− 1.7	+ .216 P	− 1.7 − .216 P	.612 P − .48
$g\ i$	6.80	− 40.5	+ .870 P	− 40.5 − .870 P	5.140 P − 239.2
$g\ h$	20.58	− 14.5	− .270 P	− 14.5 − .270 P	1.457 P + 78.4
$h\ j$	8.94	+ 41.0	− .885 P	+ 41.0 − .885 P	7.000 P − 324.0
$h\ i$	9.51	+ 13.5	+ .136 P	+ 13.5 + .136 P	.176 P + 17.5
$i\ k$	4.50	− 22.2	+ .962 P	− 22.2 + .962 P	4.162 P − 96.8
$i\ j$	16.60	− 32.0	− .162 P	− 32.0 − .162 P	.435 P + 86.0
$j\ l$	6.04	+ 23.0	− .985 P	+ 23.0 − .985 P	5.850 P − 136.5
$j\ k$	8.60	+ 24.6	+ .066 P	+ 24.6 + .066 P	.037 P + 14.0
$k\ m$	3.68	0	+1.000 P	0 + P	3.680 P
$k\ l$	17.35	− 44.0	− .075 P	− 44.0 − .075 P	.097 P + 57.2
$l\ m$	7.25	+ 32.5	+ .284 P	+ 32.5 + .284 P	.586 P + 66.9
$l\ n$	4.70	0	−1.000 P	− P	4.70 P
$m\ o$	3.68	0	+1.000 P	+ P	3.68 P

and the whole work done in deforming the whole structure will be $U = \Sigma\,u$, and, as before, we have

$$\frac{d\,\mathrm{U}}{d\,\mathrm{P}} = 0 = \mathbf{z}\frac{d\,u}{d\,\mathrm{P}} = \mathbf{z}\,\frac{\mathrm{L\,S}}{\Omega\,\mathrm{E}} \cdot \frac{d\,\mathrm{S}}{d\,\mathrm{P}}.$$

The values of

$$\frac{\mathrm{L\,S}}{\Omega\,\mathrm{E}} \cdot \frac{d\,\mathrm{S}}{d\,\mathrm{P}}$$

will be found to be those given in the last column.

Adding up this Table and doubling the values for all the bars save $l\,n$ and $m\,o$, to allow for the right-hand span, it will be found that

$$115.34\,\mathrm{P} = 4474\,; \text{ whence } \mathrm{P} = 38.8 \text{ tons.}$$

Knowing this, the stresses in all the remaining bars can be determined. If one span only was loaded, P would obviously be half the above, or 21.4 tons. The effect of the stress P on the bars $l\,n$ and $l\,o$ is, of course, to cause an upward reaction at each of the abutments equal to

$$\frac{l\ m}{a\ m} \cdot \mathrm{P}.$$

If, as usual, the effect of the bracing bars on the deflection had been neglected, we should have had, for the case in which the load covers both spans, P = 44.5 tons, an increase of very nearly 15 per cent.

SOLID BEAMS.

The structures considered up to the present have been frames in which questions of direct tension and compression only had to be dealt with. The most numerous examples of structures with superfluous parts are, however, to be found in arches, continuous girders, and stiffened suspension bridges, in which questions of bending and shearing arise.

To the determination of the stresses in such structures the method of least work can be very advantageously applied. As a preliminary it will be necessary to find an expression for the work done in deforming a bar by shearing and bending.

The work due to shearing is in general quite negligible in practice in comparison with the stresses due to bending and direct tensions and compressions. It is, however, very easy to deduce an expression for it.

If a bar is subjected to a uniform shearing force S, two parallel sections of the bar will slide relatively to each other. Suppose, for the moment, the stress uniformly distributed over the section, and the two sections to be L inches apart. Then the distance one section will move relatively to the other will be $\dfrac{S}{\Omega} \cdot \dfrac{L}{G}$ where Ω = the area of the bar, and G is the shear modulus of elasticity. Then, if the shearing force increases from 0 to its final value S, its average value will be $\dfrac{S}{2}$, and hence the work done between the two sections will be

$$\frac{S^2 L}{2\Omega G}.$$

As a matter of fact, however, the shearing stress is seldom uniformly distributed over a section, and hence the above value for the work done has to be multiplied by a coefficient, and we may therefore write,

$$\text{Work done by shearing} = U_s = A. \frac{S^2 L}{2n.G}.$$

According to Castigliano, for cases of simple shear the coefficient $A = \frac{6}{5}$, when the section sheared is rectangular, and the depth of the section is not very great compared with the width.

For an elliptic section $A = \frac{10}{9}$ provided the ellipse is not very flat.

In a similar manner, for cases of torsional shear, the work done will be $U_s = \frac{C\ T^2\ L}{2\ J\ G}$ where J is the polar moment of inertia of the section, T the torsional moment, and C a constant depending on the form of the section. For a circle C is unity, and for a rectangle of which the breadth is not very small compared with the depth, $\frac{C}{J} = \frac{3}{b\ c^3}$ where b and c are the sides of the rectangle, and $b > c$.

Coming now to cases of bending, consider a beam E A B D (Fig. 8) supported at A and B, and fitted with scale pans at its extremities in which weights can be placed. Also suppose the weights in the two pans to be always equal. Then the bending moment curve will be as in Fig. 9, being uniform between A and B. Further, let the lengths outside the supports be very rigid, as compared with the remainder of the beam. Then, if equal quantities of shot are gradually poured into the two scale pans, the lower surface of

C

the beam will take a shape something like that
indicated in Fig. 10. The parts A D and B E will
each turn through a small angle a, whilst remaining
tangents to the under surface of the beam at A and B.
Hence perpendiculars to the beam at A and B will
make an angle $i = 2a$ with each other.

Now, during this bending of the beam, the scale
pan at each end has moved through a distance
$a \sin a$, and the average value of the load has been
$\dfrac{W}{2}$.

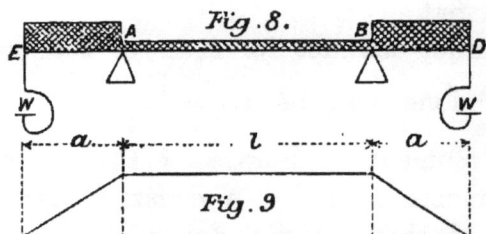

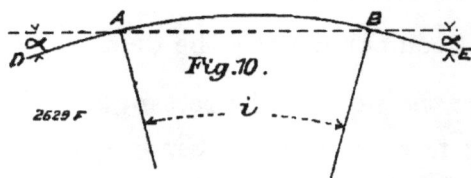

Fig. 8.

Fig. 9

Fig. 10.

Hence the total work done by the sinking of the
loads will be $W \cdot a \cdot \sin a$. But $W \cdot a =$ the bending
moment in the beam between A and B.

Therefore the work done will be $M \sin a = Ma$,
when a is small, as it always is in practice. But
$a = \dfrac{i}{2}$, therefore the work done on a given length
of a beam by a uniform bending moment is equal

to that moment multiplied by half the change in the angle between two normal sections of the beam taken at the extremities of the length considered.

Now to determine the value of i.

The stress on the upper surface of a beam is shown in ordinary works on statics to be:

$$p = \frac{M y}{I},$$

where I denotes the moment of inertia of the beam, and y the distance of the upper surface from the neutral axis. The upper fibres will be elongated by an amount

$$\lambda_1 = \frac{p\,l}{E} = \frac{M y l}{E I}.$$

Similarly the lower fibres will be compressed by an amount

$$\lambda_2 = \frac{M z l}{E I},$$

where z is the distance of the lower fibres from the neutral axis.

Then

$$y + z = h,$$

the depth of the beam, and hence the difference in length of the upper fibres and the lower will after stress be

$$\lambda_1 + \lambda_2 = \frac{M h l}{E I}.$$

But if ρ_1 be the radius of the upper surface of the beam and ρ_2 the radius of the lower, we will have

$$(\rho_1 - \rho_2)\, i = \lambda_1 + \lambda_2 = \frac{M h l}{E I}.$$

But, in the case considered above,

$$\rho_1 - \rho_2 = h \; ; \; \therefore \; i = \frac{M l}{E I}.$$

And hence finally the work done in deforming a beam by a uniform moment M will be :

$$\frac{M\,i}{2} = \frac{M^2\,l}{2\,E\,I}.$$

But in most cases which occur in practice the bending moment is not uniform. Nevertheless, we may divide up the whole beam into a number of very small parts, throughout which the bending moment may be considered uniform. Then the total work done in bending the beam may be obtained by calculating the work done on each of the parts on the above assumption, and adding the whole together. The greater the number of parts into which the beam is thus considered to be divided, the greater the accuracy of the result.

From this it follows that the work done in deforming a beam by bending is equal to the length of the beam multiplied by the average value of $\dfrac{M^2}{2\,E\,I}$.

THE DEFLECTION OF A CURVED BEAM.

The average value of $\dfrac{M^2}{I}$ can, in some few cases, be determined by the integral calculus, but more generally can only be determined by some approximate method, such as Simpson's rule, or those of Cotes or Gauss. A very convenient rule has also been given by Mr. Weddle. We shall make use of this later to determine the deflection of the davit shown in Fig. 11, under a load W applied as indicated. It is assumed that the davit is solidly *encastré* at its base, where it

is 4 in. in diameter, which thickness is maintained up
to the commencement of the curved portion, after
which it tapers down uniformly to a diameter of 2 in.
at the extreme end. To determine the deflection of
this, we have, as shown on page 21, vertical deflection

$$\Delta_v = \frac{d}{d} \frac{U}{W},$$

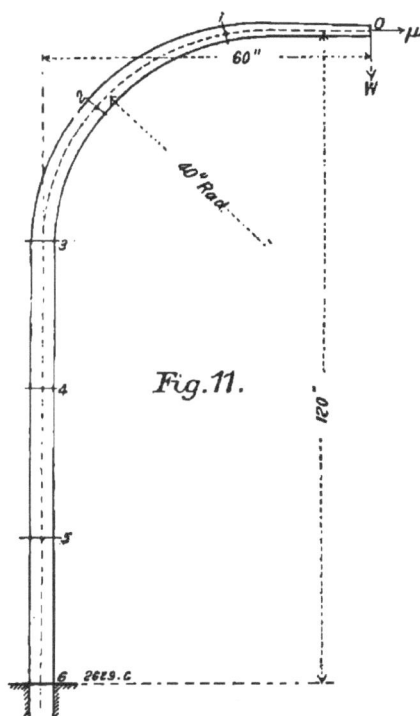

Fig. 11.

where U is the work done in deforming the beam.
But

$$U = \frac{l}{2E} \times \left(\text{mean value of } \frac{M^2}{I} \cdot \right)$$

To determine this mean value, divide the centre line

of the beam into six equal parts, as shown, and calculate the value of

$$y = \frac{M^2}{I},$$

for each of the points 0, 1, 2, 3, 4, 5, and 6. Then if y_0 is the value at 0, y_1 that at 1, and so on, we have, by Mr. Weddle's rule, mean value of

$$y = \frac{1}{20}\left[y_0 + y_2 + y_4 + y_6 + 5\,(y_1 + y_5) + 6\,y_3\right].$$

Similarly

$$\frac{d\,U}{d\,W} = \frac{l}{2\,E}\cdot\left\{\text{mean value of}\,\frac{d\,y}{d\,W}\right] = \Delta_v$$

The values of $\dfrac{d\,y}{d\,W}$ for the various sections are given below :

Section.	I.	M.	$\dfrac{M^2}{I}$	$\dfrac{d\,y}{d\,W}$
0	.7854	0	0	0
1	2.47	27.0 W	295 W²	590 W
2	6.06	50.4 W	409 W²	818 W
3	12.57	60.0 W	286 W²	572 W
4	12.57	60.0 W	286 W²	572 W
5	12.57	60.0 W	286 W²	572 W
6	12.57	60.0 W	286 W²	572 W

Then applying Weddle's rule, the average value of $\dfrac{d\,y}{d\,W}$ will be found to be 560 W,

and the deflection

$$\Delta_v = \frac{d\,W}{d\,W} = \frac{l}{2\,E}\cdot\ 560\ W$$

$$l = 163''.\quad E = 13,000 \text{ tons};$$
$$\therefore \Delta = 3.5\,W.$$

Hence if

$$W = \tfrac{1}{4} \text{ ton, } \Delta = \tfrac{7}{8}''.$$

If it were desired to determine the horizontal deflection of the end of the davit, under the load W, in addition to the vertical one, a force μ may be assumed to act horizontally as indicated in the figure. Then the moment at point 1 is now $27.0\ W + .6\ \mu$. At point 2 it is $50.4\ W + 14\ \mu$, and similarly for the other points. Then the $\dfrac{M^2}{I}$ can be found for each of these points, and

$$U = \frac{l}{2\,E} \times \left(\text{average value of } \frac{M^2}{I} \right);$$

and, finally, the horizontal deflection is equal to $\dfrac{d\ U}{d\ \mu}$, whatever the value of μ may be. Hence, if, after having obtained $\dfrac{d\ U}{d\ \mu}$, μ is made 0, the horizontal deflection of the davit end under the influence of W acting alone is determined. Using Weddle's rule, the horizontal deflection will be found to be $\Delta_H = 2.39\ W$ in. where W is the load in tons. Six sections is, perhaps, in this case, rather too small a number of parts in which to divide the beam in question, if any special accuracy were required, but is enough for the purpose of illustration. The device of using an imaginary load to determine deflections, as in the above example, and in that of the Hawkesbury Bridge, dealt with on page 24, is, we believe, original with Castigliano.

ARCHES.

Proceeding now to the discussion of some arch problems, let us suppose A C B, Fig. 12, to be the centre line of an arch rib, and let it be loaded by

a weight W as indicated.   Assume D E F to represent
the line of resistance of this load.   Then a certain
thrust T, say, acts on the arch along the line D E,

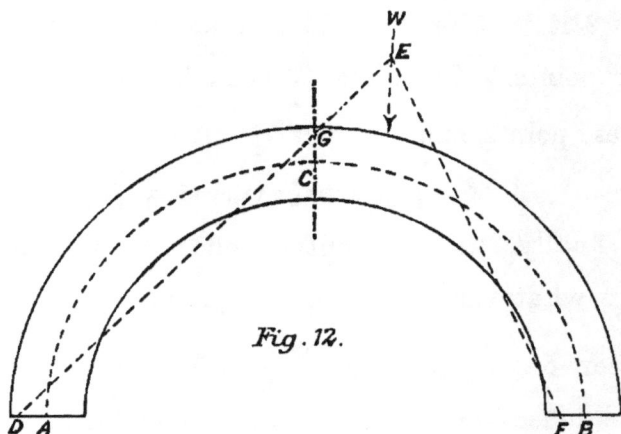

Fig. 12.

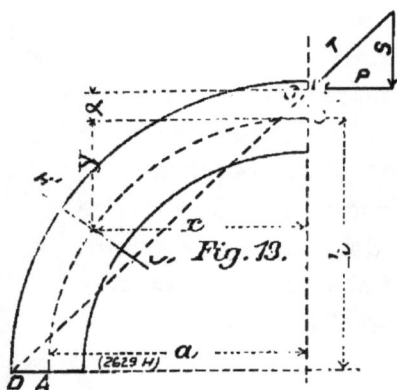

Fig. 13.

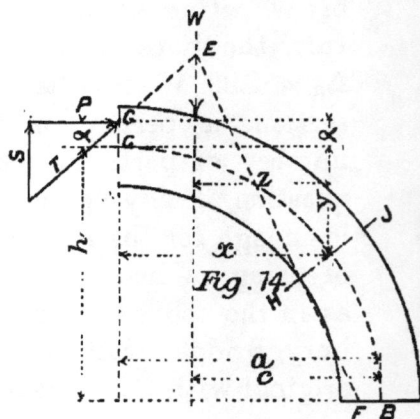

Fig. 14.

and another thrust along the line E F.   Let G be the
point at which the line D E cuts the vertical through
C.   Now the action of the left-hand part of the arch
on the right-hand part can be replaced by a force

equal to T, acting in the direction indicated in Fig. 14, whilst the action of the right-hand part of the arch on the left-hand part can be replaced by a numerically equal force T acting as indicated in Fig. 13. These forces T may be resolved into their horizontal and vertical components P and S, as shown. Now consider a section H J taken at the right-hand side of the arch. Then the bending moment at this section is evidently

$$P (a + y) + S x - W z ;$$

or if

$$P a = N,$$

the above expression will reduce to

$$N + P y + S x - W z.$$

Turning now to the left-hand portion of the arch Fig. 12, it will be seen that the bending moment at H′ J′ is

$$N + Py - S x$$

where P and S now tend to bend the beam in opposite directions. This relation will always hold, and in whatever way the arch is loaded. P and S on one half will tend to bend the arch in the same direction, and on the other half in opposite directions.

From the above it will be evident that if N, P, and S can be determined, then the bending stresses at all points of the arch can be found. Thus, if the arch is hinged at A, C, and B, the bending moment at each of these points must vanish, as the arch cannot resist bending there. In that case we should have $N = 0$, as T must pass through C.

Also taking the right-hand half, we have

$$\text{Moment at hinge B} = P h + S a - W c = 0,$$

and on the left hand

$$\text{Moment at A} = P\,h - S\,a = 0.$$

Whence

$$P = \frac{W\,c}{2\,h}$$

and

$$S = \frac{W\,c}{2\,a}.$$

If, however, there is no hinge at C, it is no longer justifiable to assume that N is 0; but there being hinges at A and B, we should have

$$N + P\,h + S\,a - W\,c = 0$$

and

$$N + P\,h - S\,a = 0.$$

Whence

$$S = \frac{W\,c}{2\,a},$$

as before, and

$$N = \frac{W\,c}{2} - P\,h.$$

Thus in this case the three unknown quantities reduce to one, and this latter cannot be determined by pure statics. Finally, if there are no hinges at A and B, it is not justifiable to assume that the moments at the abutments are zero; and hence, in this case, no one of the quantities N, P, and S can be determined by pure statics. We have, however, the general result holding for all elastic bodies, that the work done in deforming them is always the least possible consistent with the maintenance of equilibrium. If, then, in the case of the two-hinged arch we determine P so that the work done is a minimum, N can then be found from the equation

$$N = \frac{W}{2}c - P\,h.$$

In the case of the arch rigid at both abutments and crown, the whole three quantities, N, P, and S, must be chosen so as to make U a minimum.

Two-hinged Arches.

As an example of a two-hinged arch, take the case shown in Fig. 15, which represents diagrammatic-

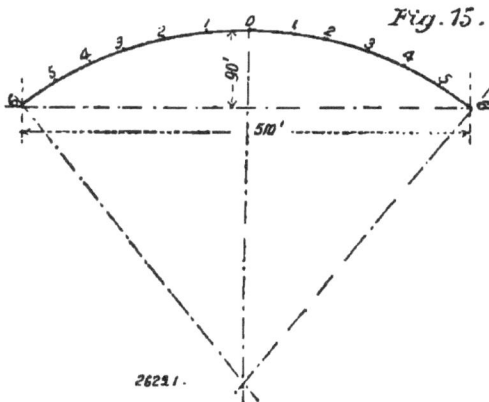

Fig. 15.

ally the Washington Bridge over the Harlem River, New York. The span in this case is 510 ft. between the centres of hinges, and the rise of the centre line is 90 ft. The dead load on an intermediate rib is said to be 1469.5 tons, and the live load 370 tons. Each rib is 13 ft. deep, and is of I-section, built up of plates and angles.

The work done by the various stresses acting on the rib can be divided under three heads, viz., that due to bending, that due to direct compression, and finally, that due to the shearing forces. As has been shown, the work done in bending a bar is

$$U_B = \frac{l}{2\,E} \times \left(\text{mean value of } \frac{M^2}{I} \right),$$

where M is the bending moment at any point of the bar, and I the corresponding moment of inertia of the cross-section ; and similarly

$$\frac{d\,U_B}{d\,P} = \frac{l}{E} \times \left[\text{mean value of} \frac{M}{I} \cdot \frac{d\,M}{d\,P} \right].$$

To find this mean value, divide up the centre line of the arch as indicated in Fig. 15, and determine the bending moment at each of the points of division.   As before, the action of the left-hand half of the arch on the other half may be replaced by a moment N, a horizontal thrust P, and a shear S.   Then, similarly to the previous instance, Fig. 14, the moment at any section, say (3), will be

$$M_3 = N + P\,y + S\,x - \omega\,\frac{x^2}{2},$$

where ω is the load per foot run of the span.

Taking lengths in inches,

$$M_3 = N + 277.9\,P + 1622.4\,S - 395,626 \text{ for the right-hand side,}$$
$$= N + 277.9\,P - 1622.4\,S - 316,049 \text{ for the left hand-side.}$$

These bending moments are given in Table IV. which has been calculated on the assumption that the live load extends over the right-hand half of the span only.   The values of I, as determined from the drawings, will be found in the fourth column. At section 6 there is a hinge, hence here the bending moment and I are both zero.

In this Table N, P, and S denote, as before, the bending moment, the horizontal thrust, and the shear at the crown, and this notation will be maintained throughout. Then at section 2, say, on the right hand, we have :

$$\frac{M}{I} \cdot \frac{d\,M}{d\,P} = \left\{ \frac{N + 124.2\,P + 1093.3\,S - 179,644}{1,364,000} \right\}$$

$$\times \left\{ \frac{d\,N}{d\,P} + 124.2 \right\} ;$$

but

$$\frac{d\,N}{d\,P} = -\,h, = -\,1080 \text{ in.}$$

h being the versed sine of the arch,

$$\therefore \frac{M}{I} \cdot \frac{d\,M}{d\,P} = \left\{ \frac{N + 124.2\,P + 1093.3\,S - 179,644}{1,364,000} \right\}$$

$$\times \left\{ -955.8 \right\}$$

Similarly at the corresponding section on the left-hand side of the arch, we should have

$$\frac{M}{I} \cdot \frac{d\,M}{d\,P} = \left\{ \frac{N + 124.2\,P - 1093.3\,S - 143,510}{1,364,000} \right\}$$

$$\times \left\{ -955.8 \right\} \cdot$$

From which it will be seen that the terms in S cancel each other when both sides of the arch are considered. Hence, neglecting terms in S, the values of

$$\frac{M}{I} \cdot \frac{d\,M}{d\,P},$$

will be found in Table V., the coefficients of S not being calculated as taking both sides into account, they cancel each other, as already mentioned.

TABLE IV.

Section.	Moment on Right-Hand Half. inch-tons.	Moment on Left-Hand Half. inch-tons.	I.
0	N	N	1,227,000
1	$N + 31.1\,P + 550.2\,S - 45{,}490$	$N + 31.1\,P - 550.2\,S - 36{,}340$	1,257,000
2	$N + 124.2\,P + 1093.3\,S - 179{,}644$	$N + 124.2\,P - 1093.3\,S - 143{,}510$	1,364,000
3	$N + 277.9\,P + 1622.4\,S - 395{,}626$	$N + 277.9\,P - 1622.4\,S - 316{,}049$	1,435,000
4	$N + 490.6\,P + 2130.9\,S - 682{,}434$	$N + 490.6\,P - 2130.9\,S - 545{,}167$	1,435,000
5	$N + 758.9\,P + 2612.1\,S - 1{,}025{,}454$	$N + 758.9\,P - 2612.1\,S - 819{,}193$	1,337,000
6	0	0	0

TABLE V.

Section.	Right-Hand Half. $\dfrac{M}{I} \cdot \dfrac{dM}{dP} = \dfrac{1}{1000} \times$	Left-Hand Half. $\dfrac{M}{I} \cdot \dfrac{dM}{dP} = \dfrac{1}{1000} \times$
0	$-.88020\,N$	$-.88020\,N$
1	$-.83445\,N - 25.951\,P - \alpha\,S + 37{,}960$	$-.83445\,N - 25.951\,P + \alpha\,S + 30{,}324$
2	$-.70073\,N - 87.031\,P - \beta\,S + 125{,}883$	$-.70073\,N - 87.031\,P + \beta\,S + 100{,}562$
3	$-.55895\,N - 155.692\,P - \gamma\,S + 220{,}628$	$-.55895\,N - 155.692\,P + \gamma\,S + 176{,}658$
4	$-.41073\,N - 201.505\,P - \delta\,S + 280{,}298$	$-.41073\,N - 201.505\,P + \delta\,S + 223{,}918$
5	$-.24016\,N - 182.261\,P - \epsilon\,S + 246{,}278$	$-.24016\,N - 182.261\,P + \epsilon\,S + 196{,}741$
6	0	0

Then by Weddle's rule already explained, the mean value of

$$\frac{M}{I} \cdot \frac{d\,M}{d\,P},$$

taking the average of both sides of the arch, will be

$$\frac{-\,.53592\,N - 113.187\,P + 141,772}{1000},$$

and therefore

$$\frac{d\,U_B}{d\,P} = \frac{-\,l}{1000\,E} \left\{ .53594\,N + 113.187\,P - 141,772 \right\}$$

To this must be added the terms due to the work done in compressing the arch U_c, and that done in shearing it, U_s, say.

If T is the normal thrust at any point of the arch, and Z the shear there,

$$U_c = \frac{l}{2\,E} \times \left(\text{mean value of } \frac{T^2}{\Omega}\right)$$

and

$$U_s = \frac{l}{2\,G} \times \left(\text{mean value of } \frac{A\,Z^2}{\Omega}\right)$$

A being a constant depending on the form of cross-section, and Ω the area of cross-section.

The work done by shear is, however, always very small compared with that due to bending, save when the bar is also twisted. U_s may, therefore, in all arch problems, be put equal to 0 without practical error. There thus remains

$$\frac{d\,U_c}{d\,P}$$

only to be determined, and this is equal to

$$\frac{l}{E} \times \left(\text{mean value of } \frac{T}{\Omega} \cdot \frac{d\,T}{d\,P}\right).$$

The values of

$$\Omega \cdot T \text{ and } \frac{T}{\Omega} \cdot \frac{d T}{d P}$$

will be found to be those given below, Tables VI. and VII. :

TABLE VI.

Section.	Ω	Right-hand Half. T.	Left-Hand Half. T.
0	257.4	P	P
1	262.4	.995 P — .118 S + 165.4	.995 P + .113 S + 132.1
2	282.4	.975 P — .224 S + 328.6	.975 P + .224 S + 262.5
3	294.9	.943 P — .333 S + 487.7	.943 P + .333 S + 389.6
4	294.9	.900 P — .437 S + 640.8	.900 P + .437 S + 511.7
5	277.4	.844 P — .536 S + 785.1	.844 P + .536 S + 627.2
6	322.5	.778 P — .628 S + 919.8	.778 P + .628 S + 734.8

TABLE VII.

Section.	Right-Hand Half. $\frac{T}{\Omega} \cdot \frac{d T}{d P} = \frac{1}{1000} \times$	Left-Hand Half. $\frac{T}{\Omega} \cdot \frac{d T}{d P} = \frac{1}{1000} \times$
0	3,886 P	3,886 P
1	3.780 P — aS + 627.2	3.780 P + aS + 500.8
2	3.367 P — bS + 1134	3.367 P + bS + 906.5
3	3.018 P — cS + 1557	3.018 P + cS + 1243
4	2.747 P — dS + 1954	2.747 P + dS + 1562
5	2.567 P — eS + 2389	2.567 P + eS + 1908
6	1.877 P — fS + 2219	1.877 P + fS + 1772

From Weddle's rule, then, the mean value of

$$\frac{T}{\Omega} \cdot \frac{d T}{d P}$$

from one end of the arch to the other is

$$\text{mean value of } \frac{T}{\Omega} \cdot \frac{d T}{d P} = \frac{3.086 \text{ P} + 1319}{1000}.$$

Hence

$$\frac{d\,U_c}{d\,P} = \frac{l}{1000\,E} \cdot \left\{ 3.086\,P + 1319 \right).$$

But

$$\frac{d\,U}{d\,P} = \frac{d\,U_B}{d\,P} + \frac{d\,U_c}{d\,P} = 0;$$
$$\therefore 0.5359\,N + 110.1\,P - 143{,}091 = 0.$$

Also, since the bending moments at the abutments are zero, we have

$$N + 1080\,P - 1{,}265{,}689 = 0$$

by eliminating S from the expression for moments at both abutments.

Solving these equations,

$$N = 32{,}500 \text{ inch tons, and } P = 1140 \text{ tons.}$$

It will be observed that, to calculate the stress arising in the arch, it was necessary to know the dimensions of its various parts, whereas the very object of finding the stresses is to determine these dimensions, and in this way a difficulty arises. Experience shows, however, that for a first approximation the work done in deforming an arch by the normal pressures is small compared with that done by bending. In the case considered it only affects the value of the horizontal thrust by about 1 per cent., hence, in a first approximation, both it, and the variation of I, may be neglected. If this is done it is possible to obtain the provisional value of the horizontal thrust P from suitable tables, such as that given on next page (Table VIII.), which shows the variation in the thrust as a continuous uniform load moves over a two-hinged arched bridge. The figures give the value of C in the equation $P = C\,w\,r$; w

D

TABLE VIII.

Angle subtended between abutments.	Value of C when load extends from one abutment over fraction of span =									
	0.1	0.2	0.3	0.4	0.5	0.6	0.7	0.8	0.9	1.0
40	.02467	.09308	.2025	.3375	.4827	.6279	.7629	.8723	.9407	.9654
60	.02312	.08970	.1923	.3204	.4619	.6035	.7316	.8342	.9008	.9239
80	.02188	.08463	.1817	.3012	.4330	.5649	.6844	.7815	.8442	.8661
100	.01975	.07840	.1670	.2769	.3977	.5185	.6284	.7170	.7757	.7954
120	.01851	.07108	.1507	.2490	.3567	.4644	.5627	.6423	.6949	.7134
140	.01632	.06287	.1323	.2178	.3112	.4045	.4900	.5594	.6060	.6223
160	.01422	.05224	.1125	.1844	.2625	.3407	.4126	.4729	.5109	.5251
180	.01188	.04412	.0916	.1496	.2123	.2749	.3329	.3804	.4126	.4245

being the weight of the load in tons per foot run, and r the mean radius of the arch in feet.

By taking the value of the horizontal thrust from this table, a provisional design of the rib can be prepared, after which the thrust should be recalculated, taking into account the variation in the moment of inertia of the rib and the work done by the tangential thrust. It should, however, be pointed out that the maximum bending moments will generally occur when the live load is distributed over from $\frac{5}{8}$ to $\frac{2}{3}$ of the span, but this of course will not affect the method of calculation which is all we are concerned with here. The value of N will usually be greatest when the live load extends over the central third or fourth of the span only.

RIGID ARCHES.

As an example of an arch fixed at the abutments, let us take the recently completed Pont Morand, Lyons. This bridge is remarkable not only for the low ratio of rise to span, viz., $\dfrac{1}{15.2}$, but for the beauty of the completed structure. The span is 67.4 metres, and the rise 4.44 metres. The ribs are of box girder section, .8 metre deep at the crown, and 1 metre deep at the abutments. The dead load on a rib is 3.38 tons per metre run, and the designed live load 1.04 tons per metre run.

Dividing as before the medial line, of one-half the rib into six equal parts its moment of inertia and area in inch units at each of the points of division are about as follows:

Section.	I [Inches]⁴.	Ω Sq. In.
0	31,960	159.5
1	32,960	159.5
2	35,140	144.0
3	37,340	146.0
4	46,220	150.0
5	61,200	170.0
6	84,000	201.0

Let us first find the central bending moment, horizontal thrust, and shear for the case in which the arch is loaded from end to end with both its dead and its live load. Then if N, P, and S denote respectively this moment, pressure, and shear, the bending moment M at the other sections will be as follows, S being zero from the symmetry of conditions.

Section.	M. In.-Tons.
0	N
1	N + 4.887 P — 2,806
2	N + 19.463 P — 11,203
3	N + 43,756 P — 25,130
4	N + 77,702 P — 55,995
5	N + 121.236 P — 69,102
6	N + 174.280 P — 98,816

From this it is easy to get the values of

$$\frac{M}{I} \cdot \frac{dM}{dN}, \quad \frac{M}{I} \cdot \frac{dM}{dP}.$$

They are as follows, the values being multiplied by 20,000 simply to avoid long decimals:

Section.	$\dfrac{20,000 \text{ M } d \text{ M}}{\text{I } d \text{ N}}$	$\dfrac{20,000 \text{ M } d \text{ M}}{\text{I } d \text{ P}}$
0	.6258 N	0
1	.6072 N + 2.955 P — 1,705	2.955 N + 14.39 P — 8,295
2	.5692 N + 11.077 P — 6,376	11.077 N + 215.59 P — 124,104
3	.5356 N + 23.437 P —13,460	23.437 N + 1025.50 P — 590,321
4	.4327 N + 33.623 P —24,230	33.623 N + 2612.60 P — 1,882,714
5	.3268 N + 39.620 P —22,526	39.620 N + 4803.49 P — 2,737,844
6	.2381 N + 41.495 P —23,528	41.495 N + 7231.79 P — 4,100,384

As in previous cases, if U_B is the work done in bending,

$$\frac{d\,U_B}{d\,N} = \frac{l}{E} \cdot \left\{ \text{mean value of } \frac{M}{I} \cdot \frac{d\,M}{d\,N} \right\}$$

and

$$\frac{d\,U_B}{d\,P} = \frac{l}{E} \cdot \left(\text{mean value of } \frac{M}{I} \cdot \frac{d\,M}{d\,P} \right)$$

Applying Weddell's formula, we get

$$20{,}000 \frac{d\,U_B}{d\,N} = \frac{l}{E} \cdot \left\{ \begin{array}{l} .4875 \text{ N} + 21.985 \text{ P} - \\ 12{,}816 \end{array} \right\} = 0. \qquad . \qquad . \qquad (1)$$

and

$$20{,}000 \frac{d\,U_B}{d\,P} = \frac{l}{E} \cdot \left[21.985 \text{ N} + 2015.1 \text{ P} - 1{,}168{,}988 \right]$$

But account has to be taken of the work U_T done in the direct compression of the rib, and a term $20{,}000\,\dfrac{d\,U_T}{d\,P}$ must be added to the last equation before equating it to zero. T is most easily determined as indicated on Fig. 16, where A B C represents the right-hand side of the arch, and O its centre of curvature. On O A take O E = 1, and on this, as

diameter, describe the semicircle E F O.   Then the
resolved part of P along the arch ring at B is

$$P \cdot \frac{O\,F}{O\,E}.$$

Hence it is only necessary to scale off O F.   Similarly the resolved part of any central shearing
force S acting along the arch ring at B is

$$S \cdot \frac{E\,F}{O\,E},$$

and, finally, the resolved part of the load acting

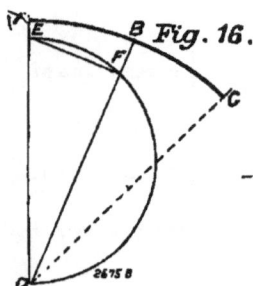

Fig. 16.

along the arch ring at B is equal to the total load
between A and B multiplied by $\frac{E\,F}{O\,E}$.   The total

thrust T is the algebraical sum of all these components,
it being noted that with unequal loading S acts
upwards on the more heavily loaded side of the arch,
and downwards on the other.   We should thus get
for section B :

$$T = \frac{O\,F}{O\,E} \cdot P + \frac{E\,F}{O\,E} \cdot \left\{\, w\,x - S \,\right\}$$

for right-hand half, and

$$T = \frac{O\,F}{O\,E} \cdot P + \frac{E\,F}{O\,E} \cdot \left[\, w^1\,x + S \,\right]$$

for left-hand half, where $w\,x =$ the total load between the crown and the section considered on the right-hand side, and $w^1 x$ the corresponding quantity on the left-hand side.

As before,

$$\frac{d\,U_T}{d\,P} = \frac{l}{\Omega}\cdot\left\{\text{mean value of}\,\frac{T}{\Omega}\cdot\frac{d\,T}{d\,P}\right\}.$$

The values of $\dfrac{T}{\Omega}\cdot\dfrac{d\,T}{d\,P}$ are as below.

Section.	Tangential Thrust. T.	$20,000\dfrac{T}{\Omega}\dfrac{d\,T}{d\,P}$.
0	P	125.38 P
1	.999 P + 1.091	125.16 P + 137
2	.996 P + 4.364	138.40 P + 607
3	.991 P + 9.790	134.52 P + 1330
4	.995 P + 21.8	129.20 P + 2860
5	.976 P + 26.9	112.06 P + 3200
6	.966 P + 38.5	92.88 P + 3702

From this, the value of $\dfrac{d\,U_T}{d\,P}$ is

$$20,000\,\frac{d\,U_T}{d\,P} = \frac{l}{E}\left[123.9\,P + 1542\right].$$

But

$$\frac{d\,U}{d\,P} = \frac{d\,U_B}{d\,P} + \frac{d\,U_T}{d\,P} = 0.$$

Therefore

$$21.985\,N + 2139.0\,P - 1,167,466 = 0.$$

Combining this with (1), and solving, it will be found that

$$N = +3126\text{ inch-tons, and }P = 514\text{ tons.}$$

If the live load only extended from one abutment up to the crown of the arch, N and P would be reduced in the ratio of the new average load on the

arch to the old one.   The average load when the
live load extends over the whole arch is 4.42 tons
per metre, and the average load when the live load
extends to the centre of the arch only is

$$\frac{3.38 + 4.42}{2} = 3.90 \text{ tons per metre.}$$

Hence the new value of

$$N = \frac{3.90}{4.42} \left(3126 \right) = 2757 \text{ inch-tons.}$$

$$P = \frac{3.90}{4.42} \left(514 \right) = 454 \text{ tons.}$$

As, however, the load is no longer symmetrical,
there will be a shearing force at the centre, the value
of which is unknown, but can be determined by the
relation $\dfrac{d\,U}{d\,S} = 0$ where S denotes this shear at
mid span.

At any section, say, for example, 3, we should have
for the bending moment on one side of the arch,

$$M = N + 43.756\,P + 669.1\,S - 25,130;$$

and at the corresponding section on the other side,

$$M = N + 43.756\,P - 669.1\,S - 19,217.$$

This gives, on the one side,

$$\frac{M}{I} \cdot \frac{d\,M}{d\,S} = \frac{669.1}{I} \cdot \left\{ N + 43.756\,P + 669.1\,S - 25,130 \right\}$$

and, on the other side,

$$\frac{M}{I} \cdot \frac{d\,M}{d\,S} = \frac{669.1}{I} \left\{ -N - 43.756\,P + 669.1\,S + 19,217 \right\} \cdot$$

Taking the mean of these, we get

Average value of $\dfrac{M}{I} \cdot \dfrac{d\,M}{d\,S}$ at section 3 =

$$\frac{669.1}{I} \left\{ 669.1\,S - 2957 \right\}.$$

Similar   results   will   be   obtained   for   the   other sections, and we thus get the figures in the second column of the annexed Table:

Section.	$\dfrac{M}{I} \cdot \dfrac{d\,M}{d\,S}.$	$\dfrac{T}{\Omega} \cdot \dfrac{d\,T}{d\,S}.$
0	0	0
1	1.517 S — 1.22	.00001 S — .00007
2	5.6789 S — 16.76	.00005 S — .00031
3	11.989 S — 52.97	.00012 S — .00103
4	17.144 S — 126.87	.00020 S — .00296
5	20.116 S — 147.38	.00027 S — .00402
6	20.957 S — 183.63	.00033 S — .00552

Again, the thrust T at the point 3, say, of the arch will on the one side be

$$.991\,P - .1302\,S + 9790,$$

and on the other

$$.991\,P + .1302\,S + 7484.$$

Then the mean value of

$$\frac{T}{\Omega} \cdot \frac{d}{d} \frac{T}{S}$$

at the two corresponding points will be

$$\frac{.1302}{\Omega} \cdot \Big\{ .1302\,S - 1.153 \Big\}, \text{ P cancelling out.}$$

The values of this quantity for the different sections are given above.

We have then

$$\frac{d\,U}{d\,S} = 0 = \frac{d\,U_B}{d\,S} + \frac{d\,U_T}{d\,S}.$$

But

$$\frac{d\,U_B}{d\,S} = l \times \text{average value of } \frac{M}{I} \cdot \frac{d\,M}{d\,S},$$

and

$$\frac{d\,U_T}{d\,S} = l \times \text{average value of } \frac{T}{\Omega} \cdot \frac{d\,T}{d\,S};$$

whence, using Weddell's rule,

$$\frac{d\,\text{U}}{d\,\text{S}} = 0 = 11.294\,\text{S} - 68.90.$$

Whence

$$\text{S} = 6.1 \text{ tons.}$$

As will be seen, the work done by S in compressing the arch is insignificant.

SUSPENSION BRIDGES.

The theory of the stiffened suspension bridge is precisely similar to that of a two-hinged arch when the stiffening girder is hinged at the abutments as usual. But whilst in an arched rib the bending and tangential forces are both resisted by a single member, the work is in the suspension bridge divided up between two separate members, one of which (the chain) resists the tensional forces only, and the other resists the whole of the bending. The theory is exactly the same. In practice, however, it is well to replace P the horizontal tension in the chain at its lowest point by q. the load per ft. which uniformly distributed over the span, would give rise to P. In other words, q is the tension in the suspenders estimated in tons per ft. run. The forces acting on the stiffening girder are then a uniformly distributed force q acting upwards, its load acting downwards, and the abutment reactions. The work done in bending the girder is then to be made a minimum with respect to q, whence the numerical value of this latter is obtained. Knowing this all the stresses in the girder are of course easily determined by pure statics.

Just as in the case of the arch to prepare a provisional design for the stiffening girder, this latter can be assumed to have a constant moment of inertia. It is then easy to show, by the method of least work, that for a continuous uniform load of p tons per foot run, advancing over the girder,

$$q = 5\,p \left\{ \frac{k^2}{2} - \frac{k^4}{2} + \frac{k^5}{5} \right\}:$$

where k is the fraction of the span covered by the load and the curve of the chain is taken as parabolic.

When the stiffening girder is fixed at the abutments, matters are somewhat more complicated, but in a similar way it can be shown that with the same assumptions as to the moment of inertia, and as to the curve of the chain.

$$q = p\,(10\,k^3 - 15\,k^4 + 6\,k^5)$$

In this case it is, however, also necessary to know the reaction at one of the abutments, and the bending moment there. If then R be the reaction at the abutment towards which the load is moving, and M the bending moment there,

$$R = W\,l\,[7\,k^4 - 4\,k^3 - 3\,k^5]$$

and

$$M = \frac{W\,l^2}{2} \left\{ k^3 - 2\,k^4 + k^5 \right]$$

where l is the length of the span in feet. When the above expression for M has a positive sign it tends to bend the girder in the same direction as the tension in the suspenders. If in place of a uniform rolling load, concentrated loads have to be considered, we have for the case of a stiffening girder hinged at its ends.

$$q = \frac{5}{l}\frac{W}{}[\lambda - 2\lambda^3 + \lambda^4]$$

where q is the upward pull of the tension rods on the stiffening girder measured in tons per foot run. $W =$ the load in tons, l the span in feet, and $\lambda\, l$ the distance of the load from one abutment. The reaction at the other abutment will then be

$$R = \frac{W}{2}[10\,\lambda^3 - 5\lambda^4 - 3\,\lambda].$$

With a suspension girder rigidly fixed at the ends we should have

$$q = 30\frac{W}{l}[\lambda^2 - 2\,\lambda^3 + \lambda^4]$$

$$R = W[28\,\lambda^3 - 12\,\lambda^2 - 15\,\lambda^4]$$

and

$$M = \frac{W\,l}{2}[3\,\lambda^2 - 8\,\lambda^3 + 5\,\lambda^4]$$

where M is the bending moment at the abutment where the reaction is R.

The three-hinged stiffening girder seems to be the favourite type, as it is much easier to calculate than the others, and there is little difficulty in taking up expansion due to changes of temperature. The central hinge is, however, badly adapted for resisting wind pressure, and the metal theoretically required in the flanges of the girder to resist deformation by a rolling load is some 4 per cent. greater than for a two-hinged stiffening girder, and about 66 per cent. more than for a stiffening girder encastré at its ends. There is also with the latter form a considerable saving in the metal required to resist wind pressure, but, on the other hand, it is more difficult to provide for expansion than with the other forms, and the stresses due to changes of temperature are considerably increased.

The above formulas of course apply, strictly speaking, only to cases in which the moment of inertia of the girder is uniform from end to end of the span. In practice, however, the variations in the value of q due to this are less than might be expected, as the fact that wind pressure has to be provided for tends to equalise the flange sections. It may, however, be useful to point out that increasing the moment of inertia at any point of a beam statically undetermined, tends to increase the bending moment at that point, and to diminish it elsewhere.

CONTINUOUS GIRDERS.

It is very easy to deduce the theorem of three moments from the principle of work, though in cases of a small number only of successive spans it is as easy to deduce the reactions directly, as was done in the case of the swing bridge already discussed, and the changes of stress to be expected from any settlement can then be ascertained somewhat more easily. We shall, therefore, content ourselves with deducing the theorem in its simplest form.

Thus let l_1 and l_2 be the lengths of two consecutive spans of a continuous girder bridge. Let M_0, M_1, and M_2 be the bending moments in the girder at the first, second, and third piers of the two spans respectively. Assume the bar cut in two at the central support, then there will be no bending in the beam there, and let the two spans be loaded in any way.

Then let m_1 denote the bending moment which would be caused by this load at any point of a simply supported beam. Then the total moment caused in the first span at any point x will be

$$M_0 \frac{l_1 - x}{l_1} - m_1.$$

Similarly the moment in the second span at any point z will be

$$\frac{M_2 z}{l_2} - m_2.$$

Suppose now the beam joined again over the central support, and let M_1 be the moment then obtaining there; then M_1 can be taken as a supernumerary moment, and hence $\dfrac{d\,U}{d\,M_1} = 0$, where $U =$ the work done in bending the girder.

The value of the bending moment at any point of the first span will now be

$$M = M_0 \left(\frac{l_1 - x}{l_1} \right) + M_1 \; \frac{x}{l_1} - m_1,$$

and at any point of the second span it will be

$$M = M_1 \frac{l_2 - z}{l_2} + M_2 \frac{z}{l_2} - m_2 ;$$

and

$$U = \frac{1}{2\,E} \int_0^{l_1} \frac{M^2}{I}\, d\,x + \frac{1}{2\,E} \int_0^{l_2} \frac{M^2}{I}\, d\,z ;$$

$$\therefore \frac{d\,U}{d\,M_1} = \frac{1}{E} \int_0^{l_1} \frac{M}{I} \frac{x}{l_1}\, d\,x + \frac{1}{E} \int_0^{l_2} \frac{M}{I} \frac{l_2 - z}{l_2} \cdot d\,z = 0 ;$$

or

$$\frac{1}{E} \cdot \int_0^{l_1} \left[M_0 \frac{l_1 - x}{l_1} + \frac{M_1 x}{l_1} - m_1 \right] \frac{x}{l_1} \cdot \frac{d\,x}{I} =$$

$$\frac{1}{E} \cdot \int_0^{l_2} \left[M_1 \frac{l_2 - z}{l_2} + M_2 \frac{z}{l_2} - m_2 \right] \frac{l_2 - z}{l_2} \cdot \frac{d\,z}{I}.$$

If I is constant, as is occasionally the case, the above expression reduces to

$$M_0\, l_1 + 2\,M_1\,(l_1 + l_2) + M_2\,l_2 =$$

$$\frac{1}{6\,l_1}\int_0^{l_1} m_1\, x\, d\,x + \frac{1}{6\,l_2}\int_0^{l_2} m_2\,(l_2 - z)\, d\,z$$

This is Clapeyron's theorem of three moments, and from it the bending moments over the successive piers of a continuous girder bridge can be deduced in turn, and finally from them the reactions. As already stated, however, it is in many cases more convenient to deduce the reactions direct, particularly where questions of settlement or deflection at the piers themselves have to be taken into account.

TROUGH FLOORS.

As an example, let us consider the case of a single line bridge floored with Lindsay's C-section troughing. The rail in such a case acts as a continuous girder resting on elastic supports. The troughing is built up of splayed channels, $\frac{3}{8}$ in. thick at the top and $\frac{1}{4}$ in. at the sides. The total depth of the corrugations of the finished floor is 7 in., and the pitch 20 in. The floor weighs 27.10 lb. per ft., and its moment of inertia is about 91.5 (inches)[4]. The span of the floor has been taken as 15 ft. between the main girders. Let us further assume that the floor, when completed, is ballasted, and the rails laid on cross sleepers at 2 ft. 6 in. apart, as indicated in Fig. 17. The tendency is constantly towards increasing the weight and stiffness of the rail, and hence for the purpose of calculation it has been assumed that this rail is an 80 lb. rail, 5 in. high, and having a moment of inertia = 31 (inches)^4.

Now if a loaded wheel rests over a sleeper, as shown in the figure, the trough under it will sink, and, owing to the stiffness of the rail, part of the load will be transferred to the other sleepers, and the amount thus transferred is easily calculable by the principle of least work.   According to this, as has been already explained, the total work done in deforming the whole structure of rails and flooring is a minimum, consistent with the equilibrium of the forces and reactions acting.   Strictly speaking we ought also to take into account the work done in deforming the main girder, but this will not sensibly modify the results, and it has, therefore, been neglected in what follows.

In the case shown in Fig. 17 it has been assumed that the rail distributes the load in varying degrees over seven sleepers spaced at 2 ft. 6 in. centres and that the trough under each sleeper acts quite independently of its neighbours, an assumption which will be justified later on.   Now consider the work done in deforming one complete section of the floor by loads placed as indicated in Fig. 18.   The distance between the rail centres has been taken as 5 ft. for convenience in calculation, though this is, of course, a shade too great for a standard gauge line.   Then in bending a beam the total work done, we have shown to be equal to $\dfrac{l}{2\,\mathrm{E}}$. × the average value of $\dfrac{\mathrm{M}^2}{\mathrm{I}}$ or ex-

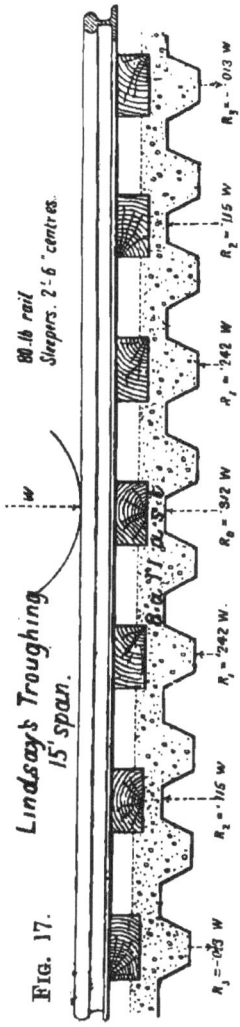

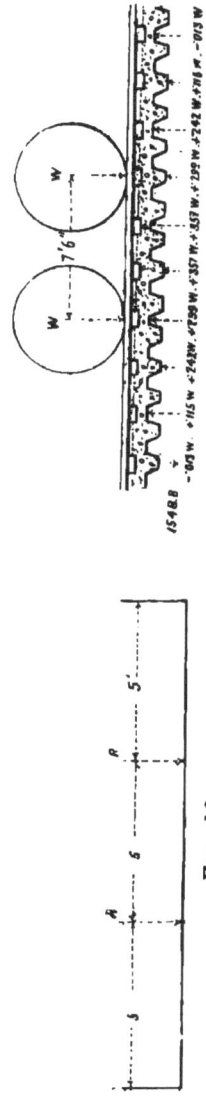

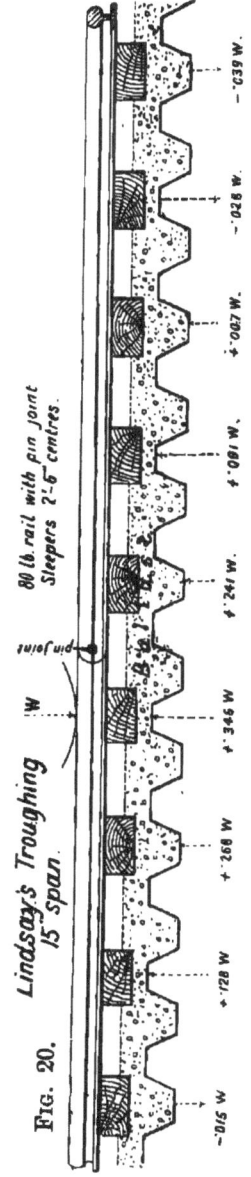

FIG. 17.

Lindsay's Troughing
15' span.

80 lb rail
Sleepers 2'-6" centres.

$R_3 = -0.13$ W
$R_2 = 1.15$ W
$R_1 = 2.42$ W
$R_0 = 3.92$ W
$R_1 = 2.42$ W
$R_2 = 1.15$ W
$R_3 = -0.13$ W

FIG. 18.

FIG. 19.

-0.13 W +1.15 W +2.42 W +3.59 W +3.57 W +3.57 W +3.59 W +2.42 W +1.15 W -0.13 W

1548B

FIG. 20.

Lindsay's Troughing
15' span.

80 lb rail with pin joint
Sleepers 2'-6" centres.

pin joint

-0.39 W
-0.28 W
+0.62 W
+1.091 W
+2.41 W
+3.46 W
+2.66 W
+1.28 W
-0.15 W

E

pressing it as an integral, we have the work done on a trough

$$= \frac{L}{2\,E\,I} \int_0^l M^2 \, dx.$$

where M = the bending moment at any point, E the elastic modulus, and I the moment of inertia of the section. Taking all units in inches, the work done in deforming one section of Lindsay's troughing 15 ft. long, by two loads each equal to R placed as shown in Fig. 18.

$$= \frac{1968}{E} \cdot R^2.$$

In the case taken there are seven reactions. That under each wheel being taken as R_0, we have for each rail, on each side of this :

Two reactions each   ...   ... $= R_1$

,,   ,,   ... $= R_2$

,,   ,, $= R_3$

And, also,

$$(R_0 + 2\,R_1 + 2\,R_2 + 2\,R_3) = W,$$

W being the load on each rail.   Hence the total work done on the troughs will be :

$$U\,\tau = \frac{1968}{E} \left\{ R_0{}^2 + 2\,R_1{}^2 + 2\,R_2{}^2 + 2\,R_3{}^2 \right\},$$

or substituting for R_0, we have

$$U\,\tau = \frac{1968}{E} \cdot [(W - 2\,R_1 - 2\,R_2 - 2\,R_3)^2 +$$
$$2\,R_1{}^2 + 2\,R_2{}^2 + 2\,R_3{}^2].$$

Similarly the work done on the rails can be ascertained in terms of R_1, R_2, and R_3.   Calling this work $U\,r$, we have the total elastic work of deformation is

$$U r + U \tau = U,$$

say, and U is a minimum, with respect to R_1, R_2, and R_3.

$$\therefore \frac{d\,U}{d\,R_1} = \frac{d\,U}{d\,R_2} = \frac{d\,U}{d\,R_3} = o.$$

It is unnecessary to go through the whole work here, but we finally get the three equations:

$$13{,}742.8\;R_3 + 8000.8\;R_2 + 5097.5\;R_1 = 1968\;W.$$
$$8{,}008.8\;R_3 + 8226.8\;R_2 + 4661.8\;R_1 = 1968\;W.$$
$$5{,}097{,}5\;R_3 + 4661.8\;R_1 + 6194.3\;R_1 = 1968\;W.$$

The solution of these equations, using seven-figure logarithms is

$$R_1 = .242.245.2\;W.$$
$$R_2 = .115.406.0\;W.$$
$$R_3 = -.013.838.8\;W.$$

Hence

$$R_0 = +.312.375.2\;W.$$

From this it appears that in the above case rather less than one-third of the weight directly over a trough is transferred to it, the rest being carried by the adjacent troughs. The results are recorded in Fig. 17. In practice, however, more than one pair of wheels has usually to be reckoned with, and if a second pair of wheels were added at a distance of 7ft. 6in. from the first pair, the resulting reactions would be those given in Fig. 19; and it is the combined effect of the two that has to be considered. In this case the maximum reaction is rather greater than one-third the load on one wheel. It will be noted that the negative reaction is less than one-seventieth of the load on one wheel.

Another question arises as to the effect of the joint in modifying the above distribution. On this head it may be remarked that with the very stiff joints now used no diminution in the distributing power of the rail is to be expected, but it may be of interest to ascertain the modification in the distribution when the rail has a perfectly flexible joint. Thus suppose the loading to be as in Fig. 20, there being a pin joint at the point shown. On making the calculation it will be found that, as was to be expected, the negative reactions are considerably increased, but the maximum positive reaction is not so much increased as might have been anticipated, the distribution being as shown in Fig. 20. As even the weakest rail joint has some degree of stiffness, it appears, therefore, that the distribution obtained in assuming a continuously stiff rail may be relied upon as that actually existing. If the load rested directly over the joint, the maximum reaction would be .443 W.

If the rails are laid directly on the troughs without the intervention of cross sleepers, a much more favourable distribution can be obtained. The distance between the points of support of the rail is now shortened to 20 in., and the maximum reaction for a single load is then .221 of the weight immediately above the trough, the distribution being as in Fig. 21. With two sets of loads 7 ft. 6 in. apart, the maximum trough reaction would be about .26 the load on one axle. This result shows the advantage of longitudinal sleepers in conjunction with trough flooring.

Coming now to the case of a double line of rail, we may take Hobson's flooring as used in the Liverpool Overhead Railway. This floor consists

of troughs pitched at 2 ft. 6 in. centres, and 15 in.
deep. It is built up of arch plates $\frac{5}{16}$ in. thick,
riveted to $\bot$-irons at the bottom. The span between
main girders is 25 ft., and the rails are carried by
longitudinal sleepers. If the rail was the typical
one already used, and fixed direct to the troughing,
we find that in the case of the inside pair of rails
the distribution would be as in Fig. 22 and for the

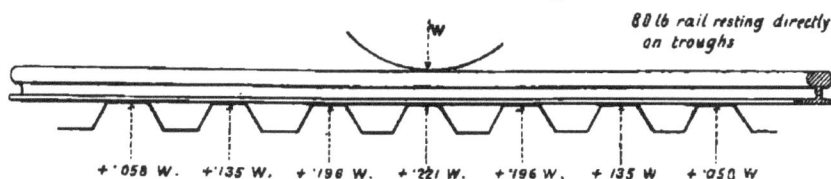

Fig. 21.

outside pair as in Fig. 23. Practically the distribution is probably a little better than this, as though
the rail actually used is not as stiff, we believe, as
the typical one taken, the combined stiffness of the
longitudinal sleeper and the rail is probably greater.
With two loads 7 ft. 6 in. apart the maximum
reactions are .34 W. and .435 W. respectively (Figs. 24
and 25). Just for comparison we have calculated out

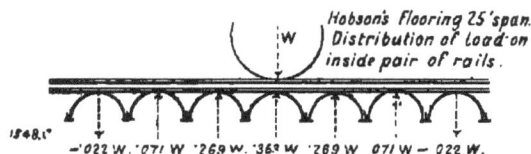

Fig. 22.

the distribution on this floor, if it had been built of
the Lindsay C section, used on the single-line spans
referred to above. The results are shown in the
diagram (Fig 24). Here it will be seen that the

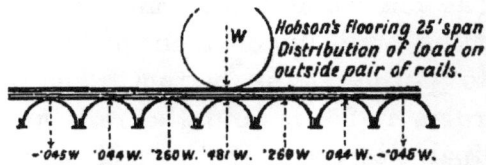

Hobson's Flooring 25' span
Distribution of load on
outside pair of rails.

−·045W. ·044W. ·260W. ·481W. ·260W. ·044W. −·045W.

FIG. 23.

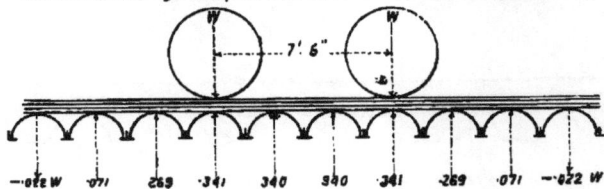

Hobson's Flooring 25' span. distribution of loads on inside rails

7' 6"

−·022W ·071 ·269 ·341 ·340 ·340 ·341 ·269 ·071 −·022 W

FIG. 24.

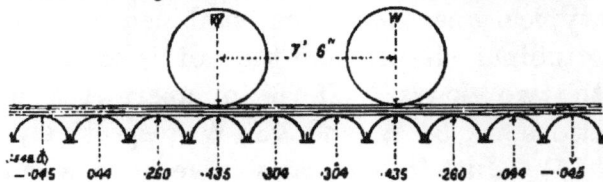

Hobson's Flooring 25' span. distribution of load on outside rails

7' 6"

−·045 ·044 ·260 ·435 ·304 ·304 ·435 ·260 ·044 −·045

FIG. 25.

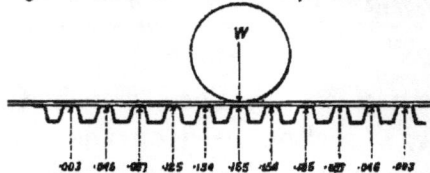

Lindsay's flooring 25 ft. span Distribution of a
single load on one of the inside pairs of rails.

·003 ·046 ·087 ·125 ·154 ·165 ·154 ·125 ·087 ·046 ·003

FIG. 2

maximum reaction is only .165 W. for a single load, and .25 W. with loads 7 ft. 6 in. apart. Hence, though Hobson's flooring, as used, has a moment of inertia of about 560 (in.)^4, as against 91.5 (in.)^4 for the light section of Lindsay's floor, taken above, yet the maximum fibre stress in Lindsay's section would, with equal loads on the centre pair of rails, be only a little more than twice as great, showing how the greater flexibility of the thinner floor has increased the distributing effect of the rails.

www.ingramcontent.com/pod-product-compliance
Lightning Source LLC
Chambersburg PA
CBHW022004190326
41519CB00010B/1377